Anatomy and Biology of Finfish

Theory Manual

THE AUTHORS

S. Shamini is an Assistant Professor in the Department of Fisheries Resource Management at the College of Fisheries Science, St. Devasahayam Institute of Fisheries Science and Technology, Midalam, Kanniyakumari dist, Tamil Nadu. She graduated with a Bachelor's degree in Fisheries Science from Dr MGR. Fisheries College and Research Institute, Ponneri, Tamilnadu and has a Master's degree in Department of Fisheries Biology and Resource Management at Fisheries College and Research Institute in Thoothukudi, Tamil Nadu. During her professional tenure, she successfully conducted a few capacity-building programmes with funding obtained from NFDB, Hyderabad. She has also published research and general articles in Fisheries Resource Management related aspects, contributed to the development of Mukkadal Museum at DIFST.

N. Karthik is an Assistant Professor in the Department of Fisheries Resource Management at the College of Fisheries Science, St. Devasahayam Institute of Fisheries Science and Technology, Midalam, Kanniyakumari dist., Tamil Nadu. He graduated with a Bachelor's degree in Fisheries Science from Dr MGR. Fisheries College and Research Institute, Ponneri, Tamil Nadu. He had his Master's degree in Fisheries Science from the Faculty of Fisheries, Sher-E-Kashmir University of Agricultural Sciences and Technology. He has qualified for ASRB – NET (National Teaching Eligibility Test) in Fisheries Resource Management (2023) and cracked the ICAR-JRF examination (2021). During his professional tenure, he diligently collected over 100 different species from the entire Kanyakumari coast, preserving them at the "Mukkadal (Tri-seas) Marine Museum" in the College of Fisheries Science, DIFST. Additionally, he has successfully conducted a capacity-building programme with funding obtained from NFDB, Hyderabad. Currently, he is leading an Institutional research project.

Prof. S. Felix is a distinguished figure in the field of fisheries education and technology consulting, with a career spanning over four decades. Currently serving as the Dean of the College of Fisheries Science at DIFST (Devasahayam Institute of Fisheries Science and Technology) in Midalam, Kanniyakumari district, Tamil Nadu and also President of AQUAFIN (Aquaculture Technology & Research Foundation India), he has held several prestigious positions, including Vice Chancellor of Tamil Nadu Fisheries University (TNJFU) and President of the World Aquaculture Society (WAS)-APC. Prof. Felix is recognized for his exemplary leadership and contributions to the fisheries and aquaculture sector, demonstrated through his skill in technologies development. He is a prolific author, having published numerous research papers, books, and manuals, and has played a pivotal role in introducing new Fisheries Science courses and establishing Fisheries Colleges and Research centers (over a dozen).

Anatomy and Biology of Finfish

Theory Manual

— Authors —

S. Shamini

N. Karthik

S. Felix

College of Fisheries Science,
St. Devasahayam Institute of
Fisheries Science & Technology,
Midalam, Kanniyakumari Dist. 629193 (Tamil Nadu)

2025

Daya Publishing House®
A Division of

Astral International Pvt. Ltd.
New Delhi – 110 002

ISBN: 978-93-5919-581-0

Published by : **Daya Publishing House®**
A Division of
Astral International Pvt. Ltd.
– ISO 9001:2015 Certified Company –
4736/23, Ansari Road, Darya Ganj
New Delhi-110 002
Ph. 011-43549197, 23278134
E-mail: info@astralint.com
Website: www.astralint.com

Printed at : **Replika Press Pvt. Ltd**

Preface

In the vast realm of Fisheries Science, understanding the intricate anatomy and biology of finfish species is fundamental. This manual serves as a comprehensive guide for students and researchers alike, delving into the external and internal structures of finfish, elucidating concepts of symmetry, and exploring the complexities of their oral region and associated systems. With a focus on practical application, it navigates through the digestive, circulatory, respiratory, excretory, and reproductive systems, shedding light on the diverse feeding habits, reproductive strategies, and developmental biology of these aquatic organisms. Moreover, it addresses critical aspects such as age determination, migration patterns, and tagging methods crucial for fisheries management and conservation efforts. Through this manual, we aim to provide a holistic understanding of finfish biology, empowering readers to contribute meaningfully to the sustainable management of aquatic resources and the preservation of biodiversity in our oceans.

S. Shamini

N. Karthik

S. Felix

Contents

Chapter-1

Study of External and Internal Anatomy of Important Groups of Finfish

Anatomy is the study of an organism's structure. Fishes come in a diverse array of forms, many with special modifications. The shape, size, and structure of body parts permit different fishes to live in different environments or in different parts of the same environment. The external anatomy of a fish can reveal a great deal about where and how it lives. Biologically speaking, a fish is made up of ten different body organ systems that work together to form the entire organism. These 10 systems cover the system by which the fish handles its food, carry away food, digest, assimilate until it is converted to waste products, the system that integrates the fish's life processes, by which it adjusts itself to the conditions of the surrounding environments, the system that provides for breathing, the system that protects against injuries, the system that supports the body and allows movement, and finally the system that works to perpetuate fish as a species.

Living species of fish are typically classified into three classes:

Agnatha: Jawless fishes, including hagfishes and lampreys. In these species, the backbone consists primarily of a notochord, which is a rod-like structure composed of unique notochordal tissues.

Chondrichthyes: Cartilaginous-skeleton fishes, such as sharks and rays. In sharks and rays, the notochord is surrounded and supported by spaced rings of cartilage.

Osteichthyes: Bony-skeleton fishes, encompassing all other living fishes. These fishes have vertebrae that form a backbone.

The study of the external and internal anatomy of these important groups of finfish provides valuable insights into their evolutionary adaptations, physiological processes, and ecological roles. This knowledge aids in understanding their behaviour, habitat preferences, and overall biology, contributing to conservation efforts and fisheries management.

External part of finfish is commonly divided in **two parts**, 1) **Head** consist of Snout, Lips, Mouth, Jaws, Teeth, Barbels, Nostril, Eyes, Operculum and Gills; 2) **Body** incudes Fins, Lateral Line, Skin and Scales.

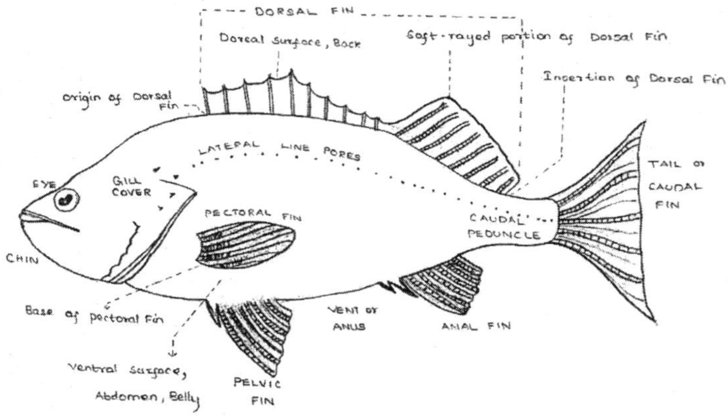

Body Forms of Fish

The shape of a fish's body is incredibly diverse and is often modified based on its habitat and lifestyle. Here are some common body forms observed in fish:

- **Torpedo-shaped (Fusiform):** This is the most common body shape among fish, characterized by a streamlined form that reduces drag as the fish moves through the water. It is often seen in fast-swimming predators like tuna and swordfish.
- **Globe-shaped (Globiform):** Some fish exhibit a round or globe-like body shape, such as pufferfishes of the family Tetraodontidae. This shape offers protection and may also be used as a defense mechanism by inflating their bodies when threatened.
- **Serpentine (Anguilliform):** Fish with a snake-like, elongated body shape belong to the Anguillidae family, which includes eels. This form allows them to navigate through tight spaces and move efficiently in complex aquatic environments.
- **Thread-like (Filiform):** Certain fish, like the snipe eel of the Nemichthyidae family, have an extremely elongated and slender body resembling a thread. This shape facilitates movement through small crevices and allows them to hunt for prey in narrow spaces.

- **Strongly Flattened from Side to Side (Compressed):** Fishes with a flattened body from side to side are adapted for life near the bottom of the ocean. Examples include butterflyfishes of the Chaetodontidae family and flounders of the Pleuronectidae family, which lie camouflaged on the ocean floor.
- **Flattened and Elongated (Trachipteriform):** Some fish have a flattened body that is greatly elongated, such as ribbon fishes of the Trachipteridae family. This shape enables them to move gracefully through the water column, often in open ocean environments.
- **Flattened from Top to Bottom (Depressed):** Skates and rays of the Rajidae family have a body shape that is flattened from top to bottom, allowing them to rest on the ocean floor and effectively blend in with their surroundings.

Body Covering and Scales in Fish

Fish have a variety of adaptations in their body covering, including skin and scales, which serve multiple functions:

Skin: The skin of a fish is typically covered by a tough outer layer. It plays a crucial role as the first line of defence against diseases and environmental factors. The skin of a fish consists of two main layers: the outer epidermis and the inner dermis or corium. The epidermis is composed of several layers of flattened, moist epithelial cells, with the deepest layers being involved in cell growth and repair. The dermal layer contains blood vessels, nerves, cutaneous sense organs, and connective tissue, playing a key role in scale formation and other integumentary structures. Additionally, the skin is involved in respiration, excretion, and osmoregulation. The diverse coloration of fish is often attributed to specialized pigment cells known as chromatophores present in the skin. Fish often have a slimy coating on their skin, which is produced by mucus cells. This coating helps reduce friction as the fish moves through water and may also provide protection against parasites and pathogens.

Scales: In fishes, scales may or may not be present. Scales range in size from microscopic to large and are an important characteristic used to classify different groups of fish. They serve as protective armor, reducing friction, and aiding in buoyancy control. The morphology and arrangement of scales can vary significantly among different species.

Agnathans: Living agnathans, or jawless fishes, are typically scaleless.

Chondrichthyans (Sharks and relatives): Sharks and their relatives have dentinal placoid scales, also known as dermal denticles, which are small, tooth-like structures embedded in the skin.

Osteichthyans (Bony fishes): Bony fishes have various types of bony scales (non-placoid scales), which can be cycloid, ctenoid, ganoid, or cosmoid, depending on the species.

<table>
<tr><th>S.No</th><th>Type of Scale</th><th colspan="2">Description</th><th>Examples</th></tr>
<tr><td>1.</td><td>Placoid</td><td colspan="2">Also known as dermal denticles. - Covered by an ectodermal cap resembling enamel (termed “vitrodentine”). - Below the vitrodentine is a thicker layer of dentine with a pulp cavity and dentinal tubules. - Each scale has a disc-like basal plate in the dermis with a cusp projecting outward through the epidermis. - Found in sharks and their relatives.</td><td>Sharks and rays</td></tr>
<tr><td rowspan="3">2.</td><td rowspan="3">Non-Placoid</td><td>Cosmoid Scale</td><td>Thinner, harder outer layer compared to placoid scales. - Covered by vitrodentine. - Below vitrodentine is a non-cellular layer called cosmine. - Beneath cosmine is a vascularized mid-layer of perforated bony substance (iropedine). - Growth occurs at the edge and beneath the scale, with no living cell layer covering it. - Found in coelacanths (Latimeria).</td><td>Coelacanths (Latimeria)</td></tr>
<tr><td>Ganoid Scale</td><td>Outer layer made of hard inorganic substance called ganoine (different from vitrodentine). - Below ganoine is a cosmine-like layer. - Innermost layer is a bony layer called isopedine. - Growth occurs at the edge, beneath, and on the surface of the scale. - Found in gars (Lepisosteidae).</td><td>Gars (Lepisosteidae)</td></tr>
<tr><td>Bony-Ridged Scale</td><td>Cycloid - it is typically smooth disc like and more or less circular mostly found in soft rayed fishes.

Ctenoid - the posterior surface or margin is toothed or comb like and a characteristic of spiny rayed bony fishes.

Typically thin and translucent. - Lacks dense enameled and dentinal layers. - Outer layer marked with bony ridges alternating with depressions. - Inner part made up of layers of criss-crossing fibrous connective tissues. - Ridges and circuli reflect changes in growth pattern, useful for identifying breeding and yearmarks. - Ctenoid scales have teeth on posterior margins.</td><td>Various fish species, e.g., salmon, trout</td></tr>
</table>

Fish Fin Types

Fish fins are classified according to their functions and positions. Fins are broadly classified into two types. Dorsal, caudal, and anal fins are the median fins (unpaired). pectoral and pelvic fins are the paired fins.

Median Fins (Unpaired Fins)	Description
Dorsal fin	Located along the back
Caudal fin	Located at the tail
Anal fin	Located at the lower edge of the body just behind the vent
Adipose fin	Fatty fin found in trouts (Salmonidae)
Modified Anal fin	Some fish have it modified into a gonopodium for copulation (e.g., live bearers - Poecilidae)

Paired Fins	Description
Pectoral fins	Supported by the pectoral girdle; absent in lampreys (Petromyzoniidae) and hagfishes (Myxinidae); enlarged in flying fishes (Exocoetidae); serve as a holdfast organ in some cling fishes (Gobiesocidae) and asiatic suckerbelly loaches (Gastromyzan)
Pelvic (ventral) fins	Depending on the species, the pelvic fins can be found in a mid-ventral (abdominal) position, underneath or just behind the pectoral fins (thoracic position), or in front of the pectorals in the throat region (jugular position).

Internal Anatomy of Finfish

The internal anatomy of finfish is a complex network of organs. As an instance, the respiratory system contains gills that extract oxygen from water, the circulatory system efficiently delivers oxygen and nutrients throughout the body. The skeletal system offers support and protection, whereas the muscular system allows movement and manoeuvrability.

In addition, the digestive system consumes food, the excretory system eliminates waste, and the reproductive system promotes species survival. The neurological system organises biological functions and responds to stimuli, whereas the endocrine system controls a variety of physiological processes through hormone production. These systems work together to ensure that the fish survives and functions properly in its environment.

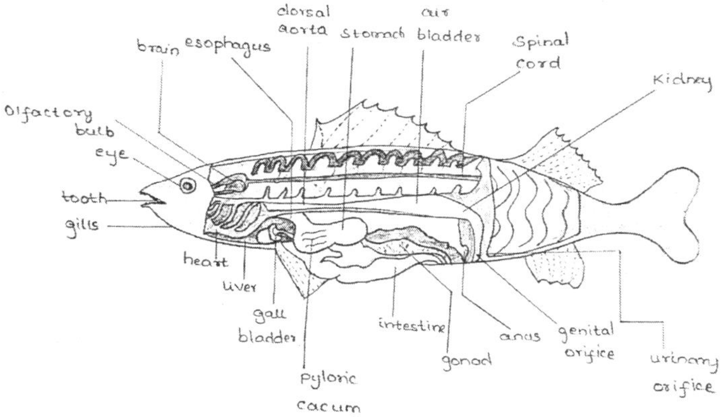

- **Oesophagus:** part of the digestive tract connecting the mouth to the stomach.
- **Stomach:** part of the digestive tract connecting the oesophagus to the intestine.
- **Intestine:** part of the digestive tract between stomach and anus.
- **Liver:** Produces bile to aid in digestion.
- **Pancreas:** Produces enzymes for digestion.
- **Gallbladder:** Stores bile produced by the liver.
- **Gills:** Used for breathing, extracting oxygen from water.
- **Heart:** Pumps blood throughout the body.
- **Dorsal aorta:** vessel in the back that carries blood from the heart to the organs.
- **Brain:** seat of the mental faculties of a fish.
- **Spinal cord:** part of the nervous system that connects the brain to all other parts.
- **Air bladder:** pocket in which urine collects.
- **Kidney**: Filter waste from the blood and regulate electrolyte balance.
- **Urinary orifice:** opening for eliminating urine.

I. Division in Animal Bodies based on Symmetry

Symmetry in animals refers to the organization of body parts in relation to a median line, exhibiting similarities in size, shape, and number across opposite sides. It enables the division of an animal body into similar halves by one or more planes.

1. Types of Symmetry

(i) Asymmetry: Some organisms, like amoebas and snails, lack any plane of symmetry, rendering them asymmetrical.

(ii) Spherical (Universal) Symmetry:

- Body can be divided into similar halves by any plane passing through the center.

  Example: Sea urchin

(iii) Radial Symmetry:

- Body consists of similar parts radiating out from a central axis.
- Can be divided into equal halves by any plane passing through the middle from top to bottom.

  Examples: Sponges, most coelenterates

(iv) Biradial Symmetry:

- Some animals exhibit radial symmetry with certain parts that are single or paired, allowing division into similar halves only by specific planes.

  Examples: Comb-jellies, sea anemones, sea stars

(v) Bilateral Symmetry

- Found in most metazoans where chief organs are paired and arranged on either side of a central axis.
- Body can be Divided into two Similar Halves by One Plane.

II. Oral Region and Associated Structure

Adaptation for Feeding

The diversity in feeding habits that fishes exhibit is the result of evolution leading to structural adaptation for getting food from the environments.

The structural adaptation of feeding organs such as

- Position and shape of the mouth
- Presence or absence of "Teeth"
- Gap of the mouth

(i) Modification in the Shape of Mouth:

- The buccal cavity is supported anteriorly by the jaws and laterally by the cheeks.
- The mouth of the majority of fishes is located at the terminal or very close to the front end of the head (subterminal).
- Bottom feeders, such as suckers and sturgeons, have subterminal or inferior mouth.
- Superior mouth observed in lie-in-wait predators.
- Leiognathids can protrude their mouths far enough to syphon in prey.
- In half beaks, the lower jaw extends into a beak, and the mouth opens above it.
- Fishes like trumpet fishes (Aulostomidae), cornet fishes (Fistularidae), pipe fishes (Syngnathidae), and butterfly fishes (Chaetodontidae) have mouths resembling elongated beaks. This adaptation is achieved through the protraction of the hyomandibular bone rather than lengthening of the lower jaw bones.
- **Feeding Methods:** These fishes may employ suction feeding or selective grazing with sharp teeth.
- **Peculiar Structure:** Half beaks (Hemiramphidae) have a unique mouth modification where the lower jaw projects into a beak, typically used for surface feeding.

(ii) Modification in the Shape of Lips:

Suctorial Feeders: Some fish species, including sturgeons (Acipenseridae) and suckers (Catastomidae), have an fleshy lip modifications. Lips of these species are mobile and described as "plicate" (having folds) or papillose (having small tufts of skin).

Barbells: Many suctorial feeders also possess well-developed barbells bordering the mouth.

Examples: The Redhorse (Moxostoma) has suctorial lips that can serve as holdfast organs in fast-flowing streams. The loach-like Gyriocheilid (Gyrinocheilus) exhibits an extremely specialized suctorial adaptation, including separate opercular structures for respiration.

(iii) Teeth:

Teeth are a crucial oral adaptation for feeding in fishes. Not all fishes bear teeth. the teeth are present in some species ; while in some they are completely disappeared.

- **Monodont -** only one type of teeth is present.
- **Heterodont -** more than one type of teeth are present.
- **Origin**: They are thought to have originated from scales covering the lips, as seen in living sharks (Squaliformes) where placoid scales transition into teeth on the jaws.
- **Types and Locations**: Teeth in bony fishes (Osteichthyes) are found in the jaw, mouth, and pharynx, including jaw teeth, roof of the mouth teeth, tongue teeth, and pharyngeal teeth on gill arch elements.
- **Variations:** Carp and suckers have teeth deep in the pharynx, while predatory fishes like Clarius and Labeo have specialized teeth for grasping, tearing, grinding, and cutting.

Type of Jaw Teeth	Description	Examples
Cardiform	Numerous, short, fine, pointed	American catfish (Ictaluridae), perches (Percidae), many sea basses (Serranidae)
Villiform	More or less elongated ' cardiform teeth	Needlefishes (Belonidae), Lionfishes (Pterois)
Canine	Elongated, subconical, straight or curved, adapted for piercing and holding	Walleyes (Alaska Pollock), Morays (Muraenidae)
Incisor	Sharp-edged cutting teeth, may fuse together forming a cutting beak	Parrotfishes (Scaridae)
Molariform	Flat with protruding denticles, for crushing and grinding	Bottom dwelling fishes like skates and rays, some Sciaenidae (Drums)

(iv) Modification in Gill Rakers:

Gill rakers serve multiple functions in fish, including protection of delicate gill filaments and specialization in food acquisition. They exhibit various forms and structures depending on the feeding habits of the fish species.

Herbivorous Fish

Gill rakers are broad sieve like structures across the gill slits for filtering the water to retain the food in the buccopharynx.

Examples: *Labeo rohita, Cirrhinus mrigala*

Carnivorous Fish

Gill rakers are long, hard and teeth – like

Example : Freshwater shark, Murrel

Omnivores

Stubby, short, and thick; unornamented.

Examples: *Puntius sarana*

Plankton Feeders

Elongated, numerous, lamellate, or ornamented to enhance filtering efficiency.

Examples: Various plankton-feeding species.

Specialized Forms

Simple but numerous gill rakers. E.g : Gizzard shads (*Dorosoma*), paddlefish (*Polyodon*).

Gill rakers resembling feathers with a main axis and branched lateral processes. *E.g:* Flatfish such as flounders and halibuts.

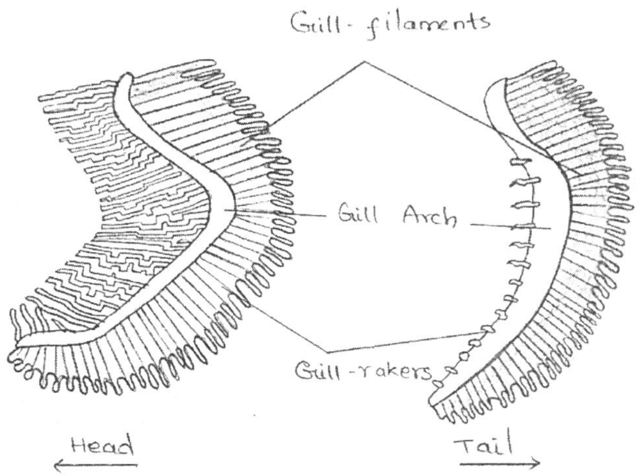

Chapter-2

Digestive System and Associated Digestive Glands

Digestion: Process by which food materials ingested by the animals are broken down into chemically simpler forms that can be readily absorbed and assimilated by the body.

Digestive System of Finfish

- It is the system of organs involved in the process of digestion.
- It is found in the coelomic cavity filled with coelomic fluid.
- The cavity is divided into two; Pericardial chamber (heart) & Peritoneal chamber (Digestive system).
- These chambers are separated by transverse septum. It has a pore which is called Pericardio peritoneal connection.
- The anatomy and physiology of the digestive system are closely related to the diet and trophic habits of the species.
- In fish, the digestive system consists of a digestive tube and associated accessory glands.
- The digestive system of fish includes the mouth, Oesophagus, stomach, pylorus, intestine, Rectum, anus and digestive glands liver and pancreas.
- Variations of the morphological pattern are exposed mainly in the shape of the mouth and the type of gill rakers, the presence of the mouth and the length of the intestine.

Digestive tract (or) Alimentary Canal

A tube connected with digestion and absorption of food. It has an opening (mouth) into which food is taken and another (anus) from which unassimilated materials is ejected. It consists of tubular and non-tubular parts. Tubular parts include Oesophagus, stomach, intestine and rectum. Non-tubular parts include Buccopharynx, Buccopharynx includes lips, buccal cavity and pharynx.

Four Zones of Alimentary Canal

- ❑ Ingressive Zone (mouth)
- ❑ Progressive Zone (Pharynx,oesophagus,stomach)
- ❑ Degressive Zone (Intestine)
- ❑ Egressive Zone (Rectum, Anus).

Component	Description
Mouth	Anterior opening of the alimentary canal. - Site of food ingestion and initial digestion. - Mouth structures vary according to feeding behavior, e.g., terminal or subterminal in different fish species.
Gill Rakers	Located in the pharynx, consisting of lateral gill slits with branchial arches. - Functions: protection of gill filaments and assistance in ingestion. - Morphology varies with diet: long and teeth-like in carnivores, short in omnivores, forming a broad sieve in herbivores, and long and thin in plankton feeders.
Oesophagus	Muscular tube connecting pharynx to stomach. - Functions: transfer of food via wave-like contractions and mucus production. - Modifications based on diet: short and narrow in herbivores/omnivores, longer and distensible in carnivores/predators.
Stomach	Part of alimentary canal between oesophagus and intestine. - Shapes vary: straight, U- or J-shaped, Y-shaped, or absent. - Functions: temporary food storage, mixing, and primary digestion to form chyme. - Modifications include thick mucosal folds in carnivores and gizzard-like structure in plankton feeders.
Intestines	Coiled tube between stomach and anus. - Continues digestion process after stomach. - Length and morphology vary based on diet: short and straight in some, long and folded/looped in others. - Complete digestion via mechanical and chemical actions with contributions from bile, pancreatic juice, and glandular secretions. Modifications based on diet: short and narrow in Carnivores, longer and coiled in herbivores and intermediate length in omnivores.
Pyloric Caeca	Muscular valve at junction of stomach and intestine. - Functions: secretion of enzymes, extension of intestines for increased surface area, and potential temporary food reservoir. - Presence varies across species.
Rectum	Terminal portion of the intestine, leading to anus. - Typically indistinguishable externally, marked by an ileo-rectal valve in some species.
Anus	Terminal opening for waste expulsion from the body.

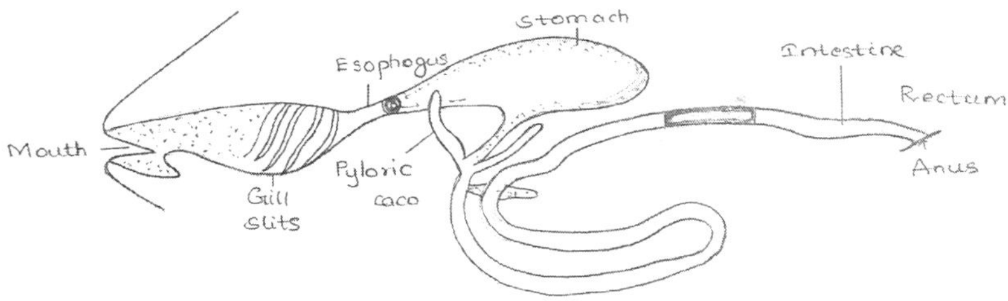

Digestive Glands

The digestive glands play vital roles in digestion, secreting fluids containing enzymes and other substances essential for breaking down food components.

1. Liver

Description: Largest gland in fish, typically bilobed with the left lobe larger than the right. Variable in size, shape, and color.

Functions:

- Secretes bile, aiding in lipid emulsification in the gut.
- Stores lipids, glycogen, vitamins A and D, and nitrogenous compounds.
- May store large amounts of lipids for buoyancy maintenance, especially in pelagic sharks.

Gall Bladder: Thin-walled, green-colored organ situated between the liver lobes, serving as a temporary storage site for bile secreted by the liver.

2. Pancreas

Description: Variable morphology among bony fishes, lacking a discrete pancreas as seen in land animals.

Distribution

- Diffused, scattered, and embedded in mesenteries, liver, clustered around the bile duct, or found in combinations of sites.
- Some species have a distinct pancreas, while others have diffused pancreatic tissue around the liver.

Functions

- Produces insulin and digestive secretions, including proteases and bicarbonates.

- Insulin stimulates amino acid uptake from the intestine and may promote growth.

Digestive Enzymes

- Digestive enzymes include pepsin, trypsin, chymotrypsin, elastase, carboxypeptidase, aminopeptidase, pancreatic alpha-amylase, colipase, and pancreatic lipase.
- Trypsin and chymotrypsin hydrolyze proteins, while pancreatic lipase breaks down triglycerides into monoglycerides and fatty acids.
- Glucosidases such as pancreatic amylase hydrolyze complex carbohydrates like glycogen and starch.

Chapter-3

Food and Feeding Habits of Finfish

Categories of Food

Food available in the nature can be classified into four categories: plankton, nekton, benthos and detritus.

Food Category	Description	Examples
Plankton	Microscopic organisms drifting with water currents. Divided into phytoplankton (with chlorophyll) and zooplankton (without chlorophyll).	Phytoplankton: Diatoms, Dinoflagellates, Zooplankton: Copepods, Crustacean larvae
Nekton	Actively swimming organisms capable of independent movement, found in pelagic (surface) or demersal (bottom) zones of water.	Fishes, Shrimps, Cuttlefishes, Squids
Benthos	Bottom dwellers with little to no movement, divided into sessile and sedentary types.	Phytobenthos: Annelid worms, Bivalves, Zoobenthos: Gastropods, Other invertebrates
Detritus	Dead and decayed organic matter of both plant and animal origin, associated with microbial flora and fecal matter.	Decomposing bits of leaves, twigs, barks, water plants, animals
Classification of Food based on Importance:		
Main or Basic Food	Preferred food consumed regularly by the fish.	**Herbivorous:** *Labeo fimbriatus, L. Rohita, Ctenopharyngodon idella, Oreochromis mossambicus*

Occasional Food	Consumed temporarily when main food is unavailable	**Omnivorous:** *Etroplus suratensis, Cyprinus carpio, Tor putitora, Cirrhinus mrigala, Clarias batrachus, Heteropneustes fossilis*
Incidental Food	Occasionally enters the gut along with main food items	-
Emergency Food	Consumed in the absence of basic food to maintain physiological activities	-

Classification of Fishes based on Food Preference		
Herbivorous	Feeds on unicellular algae, filamentous algae and higher plants, constituting at least 75% of its diet.	*Labeo fimbriatus, L. Rohita, Ctenopharyngodon idella, Oreochromis mossambicus*
Detritivorous	Feeds mainly on detritus	*Labeo calbasu, Mugil cephalus*
Omnivorous	Feeds on both plant and animal matter	*Etroplus suratensis, Cyprinus carpio, Tor putitora, Cirrhinus mrigala, Clarias batrachus, Heteropneustes fossilis*
Planktivorous	Feeds mainly on plankton (phyto and zooplankton)	Catla (zooplankton feeder), Silver carp (phytoplankton feeder)
Carnivorous	Feeds on prey organisms	*Wallago attu, Mystus singhala, Channa striatus*
Insectivorous	Feeds mainly on insects	Trout
Carcinovorous	Feeds mainly on crustaceans	Black bass
Malacovorous	Feeds mainly on molluscs like snails/clams	Black carp
Piscivorous	Feeds on fish other than its own species, often preying on small fishes of other species.	Barracuda
Larvivorous	Feeds mainly on insect and crustacean larvae/ fish larvae	*Gambusia affinis*
Cannibalistic	Feeds on the young ones of the same species	*Channa marulius, Lates calcarifer*

Classification of Fishes Based on Position in Water Column

Depending on the position in the water column they occupy, the fishes are categorized as:

- **Surface feeders** : fishes which live in the surface and feeding mostly on the food available in the uppermost layer of water. e.g. Catla feeding on zooplankton and silver carp feeding on phytoplankton.
- **Mid-water or column feeder** : fishes which live in the mid water or column water and feeding in the middle layer are called mid-water or column feeders. e.g. Rohu feeds on plant matter including decaying vegetation.
- **Bottom feeders** : fishes which live at the bottom and feed on the benthic fauna. e.g. Mrigal feeds on detritus and decaying vegetation and common carp on molluscs.

Major Fish Feeding Types

(i) Predator

- Predator fish are the one which feed on macroscopic animals.
- These fishes have well developed teeth for grasping and holding to seize their prey organism firmly.
- The predatory fishes have large mouth, reduced gill rakers, fairly large stomach and short intestine.
- The stomach is provided with acid secretions to digest the proteinaceous food material. (e.g. Ribbon fish, seer fish, barracuda etc.).
- Normally predators feed during the day time and while requires acute vision. But in case of deep sea predators, eyes are not well developed, so they locate their prey by smell, taste, touch and by lateral line sensory system and chase with large gaping mouth.

(ii) Grazers

- Grazing fishes take their food by bites on a large spread of food organisms.
- The grazers normally go in groups i.e. Butterfly fishes (chaetodontidae) and parrot fishes (sparidae) which are seen in coral reefs, feeding on coral polyps and on algae among the reef.
- Lepidophagus grazing is a special type of grazing where the fishes pluck the scales from other fishes. i.e. Cichlids.
- In hatcheries some fishes exhibit strange type of graze-feeding behaviour. i.e. Trout and salmon tear off body parts from other fishes especially fins. In such fishes, mouth is pointed and provided with plucking incisors.

(iii) Food Strainers

- They strain or sieve water to get the food material for which their gill rakers are modified.

- The gill rakers are numerous, elongated and closely set in. With the help of such gill rakes, they filter large volume of water to obtain small sized planktonic organisms.
- In plankton feeders, the food is selected by size not by kind or type i.e. Sardine, silver carp and mackerel feed on copepods, but they may take other plankton of the size of the copepods.

(iv) Food Suckers

- This type of feeding habits is seen in the case of bottom dwelling fishes.
- Here the fishes swallow their food by sucking the desired food or food containing material. In such fishes, lips are modified for sucking purpose.
- They suck food by either segregating the food before sucking in or ingest along with unwanted materials into the mouth and segregate it inside the mouth *e.g.* Sturgeon.
- On account of sucking habit many demersal fishes accumulate unwanted food material in large quantities in the stomach *e.g.* Catfish

(v) Parasitic Feeding

- Parasitism is a specialized mode of feeding and fishes are no exception to this highly evolved feeding habit.
- The parasitic fishes, feed on the body fluids of the host fish. Here the mouth is well developed and adapted to hold on to the host and pierce and suck the body fluid.

e.g. Lampreys and hag fishes have terminal mouth which is surrounded by short barbles to pierce the host fishes. Interestingly, in deep sea angler fish (ceratias), males which are smaller in size are parasitic on the large bodied female fish. Shortly after hatching, the male finds a female and attaches by its mouth to female body and the female responds by developing a fleshy papilla, from which the male can absorb nutrients. Once attached to the papilla, the male takes no free living food at all.

Stimuli for Feeding

Two kinds of stimuli are involved in feeding:

- Factors affecting the internal motivation or drive for feeding include season, time of day, light intensity, time and nature of last feeding, temperature and any other internal rhythms that may exist.
- Food stimuli perceived by the senses like smell, taste, sight and the lateral line system that release and control the momentary feeding act. The interaction of these two groups of factors determines when, what and how a fish will feed.

Detection of Food

- **Night feeders** : detecting the food by smell and taste e.g. Catfishes.

- **Sight feeder**: the fishes are active during the day time and their vision and position of eyes is important in the feeding habit. Sight feeders feed intensely during day but it does not mean that they don't feed at during the night.
- **Surface feeders:** like minnows, eyes are oriented upwards above the mid lateral line. This gives better view of objects at the surface of water.
- **Bottom feeders:** like cat fishes, the eyes are present below the mid lateral line and they can able to see the bottom area better. The fishes which feed during day as well as at night have well developed lateral line that helps in detection of food. Lateral line system is well developed in blind cave fishes and deep sea fishes.

Selection of Food

Not all the food materials that enter the mouth are swallowed by the fish. The final selection takes place in the mouth and pharynx especially in bottom dwelling fishes. These fishes feed on extraneous matter and it is subjected to selection inside the mouth. This selection is made by various sensory structures situated in mouth, tongue, gill rakers, epibranchial organs, and tissue surrounding the pharynx. The unwanted material is thrown out and in many instances unwanted food is also thrown out from the gill openings. In some cases, they throw the unwanted food by coughing action. Normally gill rakers, pharyngeal teeth, bristles and epibranchial organ serve as mechanical structures in retaining or rejecting the food.

Morphological and Anatomical Feeding Adaptations of Finfishes

Parts of alimentary canal	Herbivorous	Carnivorous	Omnivorous	Plankton Feeder
Buccopharynx & Teeth	Anterior part respiratory; posterior masticatory; teeth completely absent from jaws and palate; pharyngeal teeth for crushing prey (*e.g. Labeo rohita, Ctennopharngodon idella*)	Sharp, strong teeth, tooth like gill rakers; teeth in anterior region of cavity to prevent the escape of prey (*e.g. Channa striatus, Sphyraena barracuda, Saurida tumbil*)	Teeth intermediate condition i.e., in some species teeth present on jaws and palate, but less developed than carnivorous (*e.g. Clarias batrachus*)	Teeth may be completely absent in buccopharynx but some possess teeth on jaws (*e.g. Rastrelliger kanagurta, sardinella longiceps*)
Gill rakers	Broad, sieve like across gill slits for filtering water to retain food in buccopharynx. (*e.g. Labeo rohita, Ctennopharngodon idella*)	Long & hard (*Channa striatus, Wallogo attu*); But reduced to small tubercles in some like *Trichiurus* sp; absent in some (*Muraenosox* sp)	Short & stumpy (*e.g.* Mahseer - Tor tor)	Very long and thin; sieve like to retain plankton in buccal cavity *e.g. Rastrelliger kanagurta, sardinella longiceps*

Parts of alimentary canal	Herbivorous	Carnivorous	Omnivorous	Plankton Feeder
Oseophagus	Short & narrow (*e.g. Labeo rohita*)	Longer & distensible (*e.g. Sphyraena barracuda*)	Short & narrow (*Cyprinus carpio)*	Short & narrow
Stomach	True stomach not present in all species, anterior part of intestine is swollen to form sac behind oesophagus for food storage of carps	Sac like and thick walled	Same as herbivorous	Reduced in size but greatly thickened to become gizzard like
Intestine	Thin walled and highly coiled	Short & nearly straight	Intermediate condition	Phytoplankton feeder – herbivorous type; Zooplankton feeder – carnivorous type;
Pyloric caecae	Absent in stomachless fishes	Each caecum may be large in size but caecae less in numbers	Intermediate condition	Each caecum may be small and fine but caecae numerous

Food and Feeding Habits of Commercially Important FinFishes:

- **Indian oil sardine**: Predominantly phytoplankton feeder, favoring diatoms like *Fragillaria oceanica*; may also consume copepods, dinoflagellates, ostracods, larval prawn, fish eggs, and blue-green algae.
- **Lesser sardines:** Feed on a variety of phytoplankton and zooplankton including biddulphia, coscinodiscus, thallasiothrix, copepods, mysis, larvae of prawns and crabs, fish eggs, and molluscan larvae.
- **Anchovies:** Feed on phytoplankton (diatoms, dinoflagellates) and zooplankton (copepods, ostracods, amphipods); crustacean plankton comprises about 60% of their diet.
- **Mackerels:** Primarily plankton feeders, consuming zooplankton like cladocerans, ostracods, larval polychaetes, and occasionally phytoplankton; adults may feed on larval shrimps and fish.
- **Tuna:** Carnivorous, preying on crustaceans (larvae, juveniles, and adults of shrimps and crabs), cephalopods, eggs, larvae, and juveniles of small fishes.
- **Carangids:** Feed mostly on fishes like anchovies, sardines, silver bellies, squids, cuttlefishes, shrimps, and crabs. Young ones mainly consume prawns, squids, and anchovies.
- **Ribbon Fish:** Highly carnivorous, predatory feeders preferring small to medium-sized fishes and shrimps.
- **Bombay – Duck:** Carnivorous, partly cannibalistic, consuming golden anchovy, juveniles of its species, and non-penaeid prawns.

- **Silver belly:** Primarily zooplankton feeders.
- **Sciaenids:** Carnivores preying on crustaceans (especially prawns) as young, and on fishes (and other prey like molluscs, echinoderms, annelids) as adults.
- **Lizard Fish:** Young ones feed on crustaceans and small fishes; adults mainly consume prawns and fishes, with a small portion of copepods, cirripid larvae, and crustacean larval forms.
- **Pomfret:** Young ones feed on copepods, ostracods, amphipods, and larval stages of various organisms; adults consume copepods, crustaceans, polychaetes, and various larvae.
- **Flat fishes:** Diet includes benthic invertebrates, fishes, and cephalopods, with specific species preferring polychaetes, amphipods, small bivalves, and other fishes.
- **Perches:** Predatory, feeding on fishes and invertebrates such as crabs, prawns, and cephalopods; some may also consume cephalopods.
- **Elasmobranch:** Mostly carnivorous, with sharks feeding on pelagic teleosts and cephalopods, while skates and rays consume benthic organisms like crustaceans, molluscs, and smaller fishes.

Chapter-4

Gut Content Analysis

Introduction

- Direct observation on the feeding habits of a fish in its natural habitat is nearly impossible and therefore to ascertain the nature of a fish food, the best way is to examine its gut contents.
- Understanding fish feeding habits is important to know about their growth, breeding, and migration, as well as increasing the capture of commercially valuable fisheries resources. Fish diets reflect several ecological factors such as behaviour, habitat utilisation, energy intake, and interspecific interactions.
- Description of diets and feeding habits also provides the basis for understanding trophic interactions in aquatic food webs.
- Analysis of the different food items is difficult as the food items are normally found in a crushed or semi digested condition.

Types of Feeding Intensity Found in Fish

- **Gorged stomach:** A stomach in which the gut contents are full and occupy the entire stomach. The wall of stomach appears transparent and organisms inside the stomach could be seen.
- **Full stomach:** A stomach in which the food items occupying the entire cavity of the stomach.
- **¾ Full Stomach:** A stomach in which the food items occupying ¾ of the stomach

- **½ Full Stomach:** A stomach in which the food items occupying ½ of the stomach
- **¼ Full Stomach:** A stomach in which the food items occupying ¼ of the stomach
- **Trace stomach:** Very little or few organisms are present in the stomach
- **Empty stomach:** There will be no food item in the stomach. A little digested secretion may be present.
- **Regurgited stomach:** There will be no food item in the stomach. Wall of the stomach is shrunken.

Methods of Gut Content Analysis

Methods of gut contents analysis are broadly divisible into two,

- Qualitative
- Quantitative.

Quantitative methods of analysis are three types

- Numerical
- Gravimetric
- Volumetric.

Qualitative Methods

- The qualitative analysis consists of a complete identification of the organisms in the gut contents. Only with extensive experience and with the aid of good references it is possible to identify them from digested, broken and finely comminuted materials.
  i. Number of stomachs in which a particular food item (*i.e.* a particular species) as a percentage of the total number of stomachs is determined.
  ii. Frequency of occurrence (*i.e.* the number of times of occurrence) of all the food items among the stomachs examined is summed and the frequency of occurrence of each diet expressed as a percentage of the total number of specimens examined.

Stomach	Fish	Crab	Mollusc	Plant	Shrimp	Total
1	P			P	P	3
2	P	P		P		3
3	P	P			P	3
4	P		P	P		3
5	P	P			P	3
6	P	P	P		P	4
7		P	P	P		3
8	P		P		P	3
9	P	P		P	P	4
10	P			P	P	3
Total	9	6	4	6	7	32

According to (i)

Fish = Number of stomachs in which fish occurred / Total number of stomachs

= 9/10*100 = 90 %

Crab = 6/10*100 = 60 %

Mollusc = 4/10*100 = 40 %

Plant = 6/10*100 = 60 %

Shrimp = 7/10*100 = 70 %

According to (ii)

Fish = Number of stomachs in which fish occurred / Total number of specimens examined

= 9/32*100 = 28.15%

Crab = 6/32*100 = 18.75%

Mollusc = 4/32*100 = 12.50 %

Plant = 6/32*100 = 18.75%

Shrimp = 7/32*100 = 21.87 %

Uses

- This method demonstrates what organisms are being fed upon.
- Simple and easy method.
- This method holds good even when there is differential distribution of various food organisms in the water for the same reason that it is not biased by size or numbers of organisms comprising the food.
- This method is used as an indicator of inter-specific competition.
- This method is also used to illustrate the seasonal changes in diet composition.

Limitation

- It gives no information on quantities or numbers and size of food organisms.
- For instance, three organisms in a stomach, prawns, rotifers and diatoms present in the ratio of 1:200:2000 would all be treated by this method as 1:1:1 with reference to the stomach in question.

Quantitative Methods

1. Numerical Methods

The numerical methods are based on the counts of constituent items in the gut contents. The numerical methods have been adapted in different ways to assess the relative importance of food items and these can be classified under

- Number method
- Dominance method
- Volumetric method

A. Number Method

The total number of individuals of each food item is recorded and expressed as percentage of the total number of food organisms in the stomachs examined.

Stomach	Fish	Crab	Mollusc	Plant	Shrimp	Total
1	3	1	1	3	2	10
2	2	2	1	1	2	8
3	2	1	3	1	1	8
4	1	3	2	2	3	11
5	4	1	2	1	1	9
6	3	1	1	4	2	11
7	5	2	0	1	0	8
8	2	0	0	3	2	7
9	1	2	1	2	1	7
10	3	1	2	1	1	8
Total	26	14	13	19	15	87

Fish = Number of fish observed in the stomachs / Total number of food items*100

= 26/87*100 = 29.88%

Crab = 14/87*100 = 16.09%

Mollusc = 13/87*100 = 14.94%

Plant = 19/87*100 = 21.83%

Shrimp = 15/87*100 = 17.24%

Uses

- Very useful to study the food of plankton feeding fishes where the items can be counted with ease.

Limitations

- Allowance is made for the differences in size of food items.
- For plankton feeders; has limited use for other fishes.
- Counting of comminuted plant matter in the stomach of fish is impracticable and will not yield correct evaluations.
- Carnivore may consists of only one large sized fish and a couple of small larvae, the counting are of little value computations.

B. Dominance method

- Is a partial improvement of the occurrence method, the lack of consideration of the quantities of the food items present in the stomach, sought to be remedied.
- Stomach contents comprising the main bulk of the food materials present, is determined and the number of fish in which each such dominant food material is present is expressed as a percentage of the total number of fishes examined.
- Food items are good according to the no.of stomachs in which they occurred and is expressed according to their degree of their dominance

Limitations

- It gives only a very rough picture of the dietary of a fish. Items which are less dominant due to environmental reasons may escape notice.

C. Points (Numerical) Method

- It is an improvement on the numerical method where consideration is given to the bulk of the food items.
- Point are given to each item.
- Each number of points depends upon both whether the organism is very common in the stomach contents (highest no. of points) or rare (lowest no.) and upon the size of the organisms (i.e. one larger organism is counted as equal to a large no. of small ones).
- category is then allotted a no. of points and all the points gained by each food item are totalled and expressed as percentage of the total points. It is rapid, easy and requires no special apparatus; with experience, the method can be very accurate.

Points Method based on the Stomach Fullness

- The stomach is opened and is described according to the amount of food it contained as: full, ¾ full, ½ full, ¼ full, less than ¼ full, empty or trace.
- Then according to the degree of fullness, it is allotted 100, 75, 50, 25, 12 or 6 points respectively.
- Contents are then placed in a petri dish and the relative amount of each food item present is estimated visually points are allotted to each category as a result of this visual estimation.

2. Volumetric Method

It is more satisfactory method for quantitative analysis of gut contents. Volume forms a very suitable means of assessment. Suitable for herbivorous and mud feeding fishes where the numerical methods become meaningless as well as inaccurate. Types:

- Eye estimation method
- Points (volumetric) method
- Displacement method

A. Eye Estimation Method

- The contents of each sample are considered as unity, the various items being expressed in terms of percentage by volume as estimated by inspection.
- This method of analysis is subjective in nature and the investigators personal bias is likely to influence the result very greatly. This defect can be minimized to a great extent by the examination of large samples conducted over a long period

B. Point (Volumetric) Method

Instead of directly assessing the volume by sight as in the previous method, each food item in the stomach is allotted a certain number of points based on its volume.

Workers have taken into account both the size of the fish and the fullness of the stomach in the allotment of points.

The diet component with highest volume is given 16 points. Every other component is awarded 16, 8, 4, 2, 1 and 0 points depending on the volume relative to the component with the highest volume.

Percentage volumes within each subsample are calculated as:

- Percentage volume of the prey component = number of points allotted to component/total point allotted to sub sample*100
- This method is very useful for analyzing omnivorous and herbivorous where measuring volumes of microscopic organisms such as diatoms and filamentous algae are very difficult.

C. Displacement Method

- Accurate one for assessing the volume.
- Volume of each food item is measured by displacement in the graduated container such as a cylinder with the smallest possible diameter for accuracy.
- Highly Suitable for the estimation of the food of carnivorous fishes. But the differential rate of digestion of the food items may sometimes affect the accuracy of the observations.
- A knowledge of the volumes of the different size groups of the food items may be of great help in estimating the volume of the whole item form the semi digested fragments.

3. Gravimetric Method

- Total weight of all food items is determined.
- Weight of each food item is expressed as a percentage of the total weight of stomach contents. It may be done by wet weight or dry weight.
- Dry weight is determined by drying the food items in an oven at 60 – 80°C until a constant weight is obtained and then weighing the dried matter.
- Wet weight is determined by removing the excess water by removing the surface water by blotting them on tissue paper and weighing.

Fish = Weight of fish in the stomachs/Total weight of stomach contents*100

Limitations

- Dry Weight estimation is more time consuming; used where accurate determinations of caloric intake is required.
- Accurate Weighing of small quantities of food matter is extremely difficult and impracticable in studies of large collections.
- Employed only in conjunction with other methods to demonstrate seasonal variations in the intensity of feeding.

Importance of Gut Content Analysis

- Gives idea about fish feeding patterns.
- Assessment of food habits is an important aspect of fisheries management.
- Provides basis for understanding in aquatic food webs.
- Helps to estimate energy budget and understand energy transfer and tropic interactions in an ecosystem.
- Helps to develop appropriate feed for the species under culture.
- Diets of fishes represent in integration of many important ecological components that include behaviour, condition, habitat use, energy intake and inter/intra specific interactions.

Limitations of Gut Content Analysis

- A fish when captured or when preserved in formalin for study, often vomits the remains of its last meal because of the chemical shock.
- Material found in the gut, cannot be considered as food.
- Methods such as morphological (position of mouth, relative gut length, etc.) and environmental (food spectrum in the aquatic system) evidence should be calculated for verification and confirmation of the gut analysis data.

Chapter-5

Food Analysis Indices

A. Simple Indices

1. Index of Fullness

- This measured as the ratio of food weight to body weight as an index of fullness, which is very widely employed. (The ratio of corresponding volume can also be used.)
- This Index can be applied to the food in the stomach, or to that in the whole digestive tract.
- It is usually expressed as parts per 10,000

Fullness index = weight of the stomach contents/ weight of fish x 10,000

2.Index of Consumption

Some authors have used not the actual weight (or volume) of the stomach contents, but their reconstructed weight: i.e. their estimated weight at time of ingestion. When reconstructed weights are used in the formula above, the index obtained has been distinguished as the index of consumption

Consumption index = weight of fish reconstructed / weight of stomach contents x 10,000

3. Index of Selection or Forage Ratio

Most fishes have a scale of preference for the organisms in their environment, so that some are consumed in large numbers, others moderately, some not al all. A quantitative index of such differences called as the forage ratio. A study of the quantities of different organisms available to the fish is made, and also of the various items in their stomach.

Selection index = forage ratio = s/b

s = percentage representation by weight, of a food organism in the stomach

b = percentage representation of the same organism in the environment.

The lower limit for this index is 0; its upper limit is indefinitely large.

4. Index of Electivity

- It is used for comparing the feeding habits of fishes and other aquatic organisms with the availability of potential food resources in natural habitats.

  Electivity index = E = s - b/ s+b

- Index has a possible range of -1 to +1, with negative values indicating avoidance or inaccessibility of the prey item, Zero indicating random selection form the environment, and positive values indicating active selection.

5. Manly-Chesson index

- Given a variety of prey types, most fishes select some food categories over others
- Measure this selectivity, a variety of indices have been developed that incorporate measures of prey use and prey availability.
- This is good for quantifying prey preference and can be calculated for two scenarios like constant prey abundance and variable prey abundance.
  - ☆ **Constant prey abundance –** used when the number of prey eaten is very small relative to its total population or when prey is replaced as in laboratory studies.
  - ☆ **Variable prey abundance –** used when the number of prey eaten is large relative to its total population in the environment

B. Compound Indices

1. Index of Preponderance (Natarajan and Jhingran, 1961)

- This index gives a summary picture of frequency of occurrence as well as bulk of various food items. It provides a definite and measurable basis of grading the various food elements. The bulk of food items can be evaluated by 1) Numerical 2) volumetric and 3) Gravimetric methods.
- As the numerical method is not suited to the index with the frequency of occurrence it magnifies the importance of smaller organisms which may appear in enormous numbers. Therefore either volumetric or gravimetric are best to assess the food items quantitatively.

- It V_i and O_i are the volume and occurrence index of food item,

$$\textbf{Index of Preponderance } I_i = \frac{v_i o_i}{\sum v_i o_i} * 100$$

2) Index of Relative Importance (IRI)

This index is an integration of measurement of number, volume and frequency of occurrence to assist in evaluating the relationship of the various food items found in the stomach. It is calculated by summing the numerical and volumetric percentages values and multiplying with frequency of occurrence percentage value. Index of relative importance,

$$\text{Index of Relative Importance } IRI_i = (\% N_i + \% V_i)\ \% O_i$$

Where, N_i , V_i , O_i represent percentages of number, volume and frequency of occurrence .

3. Absolute Importance Index

- Method is a combination of various methods like frequency of occurrence method, numerical method, and gravimetric method (weight based) method.
- Index of various food species (items) can be determined as follows:

$$AI = \% O_i + \% N_i + \% W_i$$

Chapter-6

Circulatory System of Teleost

Introduction

It is a system by which blood supplies oxygen and nutrient material to cells and the cells give out water, carbon dioxide and nitrogenous wastes into blood. In fish, this system consists of dynamic and static components, with blood flowing continuously through the body. Fish have a closed circulatory system, with blood traveling through a network of vessels. Unlike mammals, fish typically exhibit single-cycle circulation, although double circulation is found in some species.

Circulatory System of a Fish

Heart Structure

The heart in fish is a simple muscular structure located between the posterior gill arches.

Enclosed by the pericardial membrane, it typically consists of an atrium, ventricle, sinus venosus, and bulbus arteriosus. Despite having four parts, the fish heart is considered two-chambered.

Blood Composition

Fish blood contains plasma, erythrocytes (red blood cells), leukocytes (white blood cells), and thrombocytes. Erythrocytes contain hemoglobin, facilitating oxygen transport, while leukocytes are vital for the immune system. Thrombocytes aid in blood clotting.

Blood Circulation

Blood vessels, including arteries and veins, transport blood throughout the body. Arteries carry oxygenated blood from the heart, while veins return deoxygenated blood to the heart.

Plasma Composition

Fish plasma consists of water with ions (Na+, Ca2+, K+), urea, sugars, and fatty acids dissolved in it.

Blood Cell Production

In sharks and rays, blood cells are produced in the spleen, epigonal organ, and Leydig organ.

Teleost fish primarily produce blood cells in the spleen and kidneys.

Components of the Circulatory System

Arteries

- Arteries are blood vessels responsible for carrying blood away from the heart.
- They branch into thinner arterioles, which further divide into small capillaries.
- Capillaries form an extensive network in all body tissues except cartilage and epithelium.
- Arteries, arterioles, and capillaries together constitute the arterial system.
- The arterial system includes pulmonary arteries, which transport blood to the gills, and systemic arteries, which distribute blood to all other body parts.

Veins

- Veins are blood vessels that carry blood toward the heart.
- Blood passes from capillaries into veinules, which merge to form veins.
- Veins, along with venules and capillaries, make up the venous system.
- The venous system includes pulmonary and systemic divisions, as well as portal systems.
- In a portal system, blood does not return directly to the heart but passes through an interposing organ.
- Vertebrates have various portal systems, including the hepatic portal system (in the liver), renal portal system (in the kidney), and pituitary portal system (in the pituitary gland).

Blood

- Blood comprises plasma and cellular components (blood cells or corpuscles).
- Plasma accounts for about 55% of blood volume, while corpuscles make up 45%.
- Plasma contains serum, fibrinogen (a coagulable substance), and anticoagulant heparin.
- Cellular components include red blood cells (RBCs), white blood cells (WBCs), and platelets (thrombocytes).
- RBCs, or erythrocytes, carry oxygen and help remove carbon dioxide. They contain iron, water, inorganic salts, and hemoglobin.
- Hemoglobin is a respiratory pigment composed mainly of globin (95%) and heme (5%). It transports oxygen in combination with ferrous iron.
- Fish blood exhibits antifreeze properties to prevent freezing in cold environments, facilitated by antifreeze glycoproteins.
- Hemoglobin concentrations in fish blood typically range from 7 to 10 g/100 ml.
- RBC size is larger in Elasmobranchs than in teleost fish.
- Fish have smaller blood volume than other vertebrates

	Volume of Blood
Bony fish	2 -4 ml
Lampreys	8.5 ml
Hagfish	17 ml
Elasmobranch	8 ml

Heart Structure and Function

Location and Shape

- The heart is situated within a pericardial cavity below the gills.
- It is relatively anteriorly placed compared to Elasmobranchs.
- Fish hearts, known as branchial hearts or venous hearts, primarily pump venous blood to the ventral aorta and then to the gills.
- The heart typically has an S-shaped structure and is two-chambered, comprising an atrium and a ventricle.
- It consists of four compartments: sinus venosus, atrium, ventricle, and conus arteriosus or bulbus arteriosus.
- Conus arteriosusis present in Elasmobranch and lower teleost (Acipenser)
- In highly teleost bulbus arteriosus is present
- In some species both conus and bulbus arteriosus is present (Amia)

Sinus Venosus

- Thin-walled sac that receives blood from venous vessels.
- Receives blood through ducti cuvieri and hepatic veins from the liver.
- Opens into the atrium via the sinu-atrial ostium, regulated by atrial valves.

Atrium (Auricle)

- Occupies the largest part of the dorsal pericardial region.
- Thin-walled sac, resembling an inverted pyramid.
- Large muscular chamber responsible for receiving and contracting.
- In Elasmobranchs and Teleosts, it's an undivided single chamber, while in Dipnoi, it's partially divided by an incomplete interatrial septum.
- Opens into the ventricle through the atrio-ventricular aperture, regulated by atrio-ventricular valves.

Ventricle

- In Teleosts, the ventricle can be tubular, pyramidal (triangular is shape) or sac-like.
- Large muscular chamber supported by caracoid cartilage.
- In Elasmobranchs and Teleosts, it's undivided, but in Dipnoi, it's partially divided by a muscular septum.
- Leads into the conus arteriosus in Elasmobranchs and bulbus arteriosus in Teleosts.

Bulbus Arteriosus / Conus Arteriosus

- Ventricular bulbar valves prevent backflow of blood into the ventricle.
- The length of the conus is related to the number of valves.
- Various species have different numbers of valve rows to prevent backflow.

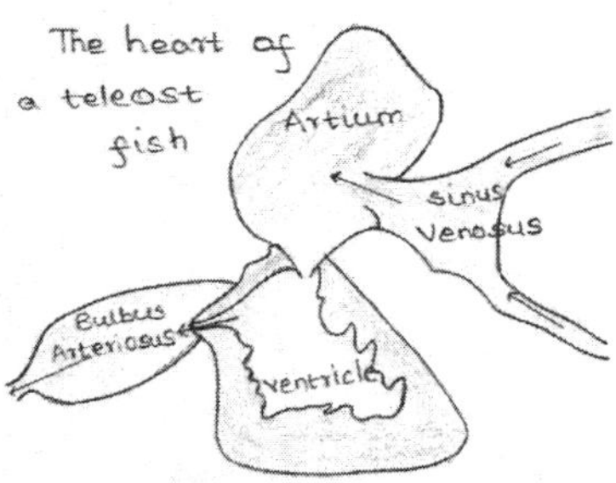

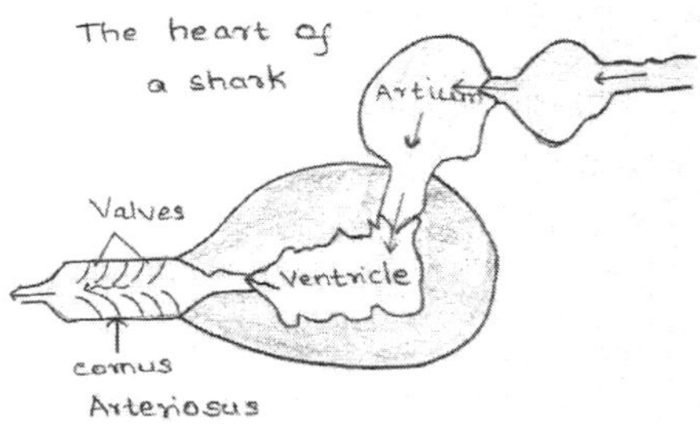

Working of the Heart

- Venous blood enters the sinus venosus and flows into the atrium, pushing apart sinu-atrial valves.
- Atrio-ventricular valves open, allowing blood to flow from the atrium to the ventricle.

- Ventricular cavity fills, causing the auriculo-ventricular valves to bulge and prevent backflow.
- Blood then moves into the bulbus arteriosus or conus arteriosus, where ventriculo-bulbar valves prevent backflow.
- Finally, blood is pumped into the ventral aorta.

Circulatory System of Lungfish

Adaptations of the Lungfish Heart

- The lungfish heart exhibits adaptations to separate the flow of blood into pulmonary and systemic circuits.
- The atrium is partially divided, with the left side receiving oxygenated blood and the right side receiving deoxygenated blood.
- Oxygenated and deoxygenated blood streams remain mostly separate as they flow through the ventricle leading to the gill arches.
- Oxygenated blood primarily goes to the anterior gill arches, while deoxygenated blood mostly goes to the posterior arches.

Structure of the Heart in Scoliodon (Sharks)

Chamber Composition

- The heart of Scoliodon (dogfish) consists of four chambers: sinus venosus, atrium (auricle), ventricle, and conus arteriosus.

Sinus Venosus

- Thin-walled chamber at the base of the pericardial cavity.
- Receives blood through large ducti cuvieri and two hepatic sinuses.
- Opens into the atrium via the sinu-auricular aperture, guarded by membranous valves to prevent backward blood flow.

Atrium and Ventricle

- Atrium lies dorsal to the ventricle and opens into it through the atrioventricular aperture, regulated by valves.
- These chambers serve as the receiving chambers for venous blood.
- The ventricle has thick, muscular, and spongy walls.

Conus Arteriosus

- The ventricle continues into the tubular conus, which extends to the anterior end of the pericardial cavity.
- Conus has two transverse rows of semi-lunar valves, each row consisting of three valves.
- Additionally, there are two small accessory valves on each side of the dorsal valve.

- The conus continues anteriorly as the ventral aorta.

Blood Flow and Functioning

- Venous blood from all parts of the body reaches the receiving chambers of the heart and is pumped to the gills for aeration.
- Rhythmic contraction and expansion of heart chambers propel blood through the heart in a definite order: Sinus → Atrium → Ventricle → Conus → Ventral aorta.
- Valves prevent backward flow of blood.
- The heart of fishes contains only venous blood, passing through the heart once during a complete circuit in the body, known as a venous or branchial heart.

Chapter-7

Respiratory System

Introduction

- In vertebrates, the respiratory system plays a crucial role in oxygen uptake and transportation to tissues. Haemoglobin, present in red blood cells, facilitates oxygen transport. Carbon dioxide, produced in tissues, is carried back to the respiratory surface for elimination.

Respiratory Organs

- Gills serve as the primary respiratory organs in fish.
- Gills possess a rich blood supply and thin epithelium covering blood vessels, maximizing surface area for gas exchange.
- Blood vessels are extensively branched and reduced to thin capillaries, facilitating close contact with the surrounding water.

Physiology of Respiration

(a) External Respiration

- External respiration involves the exchange of oxygen and carbon dioxide between blood and the surrounding environment.
- It occurs in the capillaries of the gills.
- Oxygen is absorbed by blood from the water, while carbon dioxide is released into the water.

(b) Internal Respiration

- Internal respiration, also known as tissue respiration, involves the exchange of gases between blood and tissues at the cellular level.
- Oxygen is delivered from blood to cells, while carbon dioxide produced by cellular metabolism is taken up by blood for elimination.

Gills of Teleost

In teleost, five pairs of gill arches are present of which first four bear gill lamellae, but the fifth is devoid of gill lamellae and transformed into pharyngeal bone for mastication of food. It does not play any role in respiration. Gills are lodged in gill or opercular cavities on either side of the head. They are covered by a bony plate, called Gill Operculum.

Functions of Gills

- Gas Exchange
- Osmoregulation
- Excretion
- Heat exchange
- Feeding

Functions of Gill Operculum

- To facilitate respiratory mechanism
- To protect the gills

Structure of Gills

The gill comprises Gill Arch, Gill Rakers, Gill Filaments (Primary Gill Lamellae) and Gill lamellae (Secondary Gill Lamellae)

(i) Gill Arch

- It is an important unit and bears primary (Gill Filament) and secondary lamellae.
- It encloses afferent branchial arteries and efferent branchial arteries and veins. It is also provided with nerves. It contains abductor and adductor muscles.
- Inside it contains gill rakers, taste buds, mucus gland and sensory papillae

(ii) Gill Rakers

- They are comb like structures in the inner margin of each gill arch.
- They filter the water and prevent the entry of dirty materials and food particles into the gill filaments. They act as a sieve.
- They are modified in relation to the food and feeding habits.

(iii) Gill Filaments (Primary Gill Lamellae)

- The gill filaments are blade – like structures supported by gill rays. They have reduced or no gill septum (interbranchial sptum) between two branches of filament. Active fishes generally have more filaments than sluggish fishes.
- Small bottom dwelling darters have about 300 to 500 filaments in total whereas Mackerals have 2400 and perch of size 30g have 1500.

(iv) Gill Lamellae (Secondary Gill Lamellae)

- Each gill filament is made up of secondary gill lamellae, which are actual seat of exchange of gases.
- They are generally semicircular and lined up along both the sides of the gill filaments.
- The lamellae numbers, thickness and interlamellar distance vary from species to species.
- Tuna and their close relatives have very thin lamellar walls of about 0.53 to 1.0 μm and some demersal species have 5 to 6 μm.
- Most elasmobranchs have lamellar thickness of 5 to 11 μm.
- The number of lamellae per unit length of filament also varies.
- Slow moving fishes have 10 to 20 per millimeter whereas active fishes have 30 to 80 per millimeter.

Mechanism of Respiration in Teleosts (Pumping Mechanism)

Inspiration

During inspiration, opercula and branchiostegal membrane press against the body keeping the two external branchial apertures tightly closed. Gill arches bulge laterally enlarging the internal capacity of Bucco-pharyngeal cavity, which act like a suction pump. As a result, the oral valves open and water flows in through opened mouth to fill the bucco-pharyngeal cavity.

Expiration

Now the oral valve close shutting the mouth, the gill arches contract and the opercula and branchiostegal membrane lift, opening the external branchial aperture. Consequently, water under pressure is forced to pass over gill filaments and out through the external branchial aperture.

Pumping Mechanism

The pump relies upon the principle of pressure gradient.

Water moves from high (or positive) pressure to low (or negative) pressure. The pressure gradient also opens and closes valves. There are three steps involved.

- The operculum is in its closed position after causing a negative pressure in the opercular cavity. The volume of buccal cavity increased by lowering the mouth. This produces a negative pressure in the buccal cavity causing water to be drawn into the mouth.
- The mouth is shut. The volume of buccal cavity decreases by raising the floor of the mouth causing a positive pressure in the buccal cavity. An outward movement of the operculum causes an increased negative pressure in the opercular cavity. The difference in pressure forces water through the gills.
- The operculum contracts against the body wall forcing water out of the opercular cavity past the open flaps. The cycle starts again at 1.

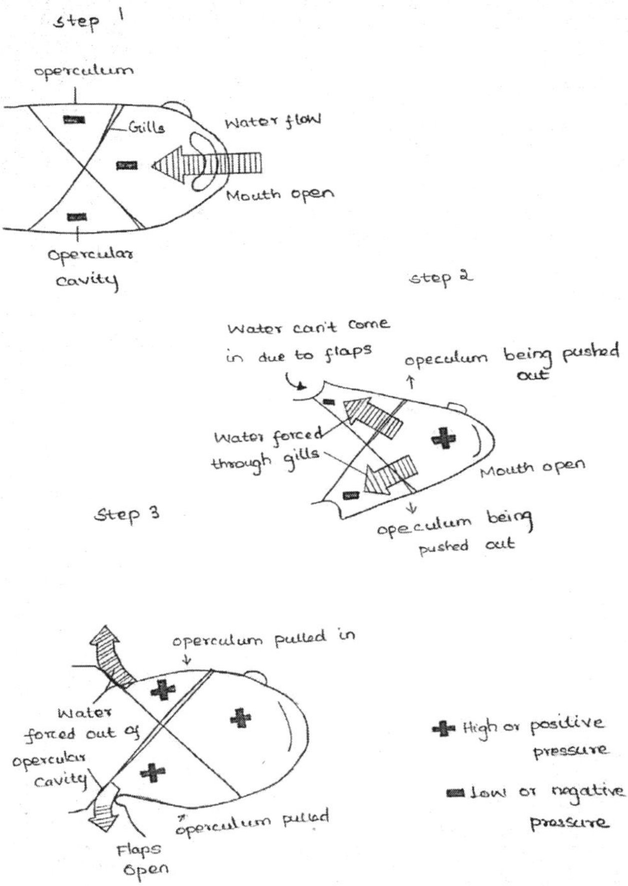

Countercurrent Flow

In most of the teleosts, blood flow in the lamellae is opposite to the water flow. Hence more oxygen diffuses in to the blood and the system is more efficient than if the flows were in same direction. This is called counter current flow.

Coughing

Process of violent sweeping of water over the gill lamellae to make them free from accumulated detritus. It is carried out by the strong abductions of the two hemibranchs of a holobranch towards each other.

Ram Ventilation

In tuna which is fast moving in nature has immobile gill cover and it require high water flow over the gill by taking water into the mantle cavity, and by ejecting the water by jet propulsion. Fast moving fishes have largest gill area than the slow movers. This help to desire more oxygen for its activity. Fast swimming fishes may leave their mouth and opercular aperture wide open. So that gills are bathed by a continuous current of water produced by swimming. This is called 'ram ventilation' The rhythmic movement of operculum is controlled by a respiratory centres which is located in the area of medulla oblongata.

Gills of Lamprey

Seven paired gill sacs or branchial pouches (gill pouches) are present in Lampreys. Each gill pouch has a thin diaphragm, which is attached to the body wall. In larval lampreys the respiratory tube has connection with mouth and oesophagus like hagfish. In the anterior vulvular velum protect the gill and regulate the water. Inside the gill pouch, the primary and secondary gill filaments are found which are the actual sites for respiration. The gill pouches are supported by elastic cartilaginous branchial basket, which is open to the outside media through the external aperture. Water is taken in by means of self contained nasal sac, which act as a hydraulic system.

Gills of Hagfish

The hagfish normally have 5-15 pairs of gill arches (6 pairs in Myxine sp. and 15 pairs in Eptatritus sp.) In Myxine a pair of common external aperture is formed by connecting 6 pairs of gill pouches each side they have one common external slit, but in primitive Hagfishes individual gill pouch opens into a separate opening.

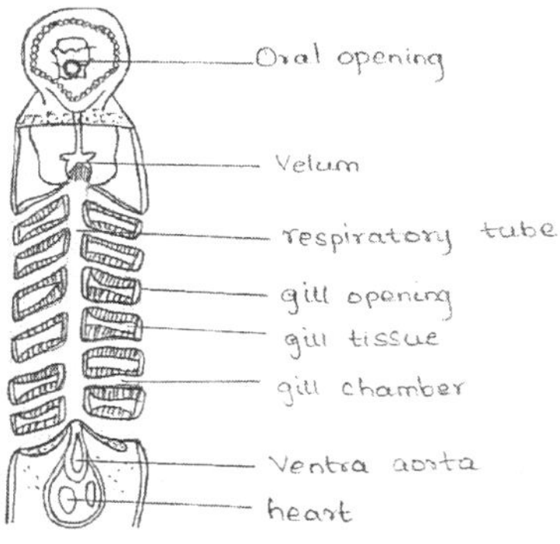

Gill of Lamprey

Gills of Shark and Rays

Five, more rarely 6 or 7 external gill slits ventrally in a series on each side in rays and their relatives, but laterally in sharks. A pair of spiracle present in elasmobranch anterior to first gill arch. There are five pairs of lateral gill pouches. Each gill pouch is compressed anterior – posteriorly and communicate with the cavity of the pharynx through a large internal branchial aperture and with the exterior through a narrow external branchial aperture called gill slits. Each gill has two sets of gill lamellae. One on its anterior wall and other on its posterior wall. Each set of lamellae is a half-gill or hemibranch so that each gill pouch has two hemibranches. The interbranchial septa between two branches of a filament (primary lamellae) extend beyond the branchial lamellae, and then each bends posteriorly to protect the lamellae. These types of gills are called Lamelliform. In sharks, respiratory water typically enters through mouth, but in the bottom dwelling rays the spiracle admits most of the water that follows sub sequentially over the gills.

Chapter-8

Accessory Respiratory Organs

The majority of teleostean fishes have well-developed gills for respiration that are highly effective at gas exchange. However, some tropical fish have evolved unique structures for assessing the respiration process known as accessory respiratory organs. These fishes are unique in their ability for bimodal mode of respiration. Generally, some existing organs have become extensively vascularized and adapted for aerial respiration. The skin, gills, branchial cavity, air bladder, and intestine have all been modified. Such accessory organs are present in species that live in shallow, stagnant water or swamps with deoxygenated water, but they rarely exist in marine fish.

Skin

- In species like *Anguilla* sp., *Amphipnous cuchia*, and *Mastacembelus*, the skin becomes highly vascular and moistened by mucus. This facilitates gas exchange between the blood and air. The skin functions as an accessory respiratory organ, especially during migration or movement over damp vegetation.

Gill

- Species like *Rita rita* and *Mastacembelus* possess unmodified gills that secrete mucus to maintain moisture. These fishes can survive for a period out of water by continuing to respire through their skin and gills.

Alimentary Canal

- In some fish species like loaches (*Misgurnus fossilis*) and Loricaridae, parts of the stomach or intestine are modified for aerial respiration. The walls become thin, transparent, and richly vascularized, allowing gas exchange.

Buccopharynx

- Fishes like *Channa* spp. and *Monopterus* spp. utilize the buccopharynx for air breathing. Air is passed through the gill slits into the opercular cavity for gas exchange.

Diverticulum of Branchial Cavity

- Species such as *Heteropneustes fossilis* and Clarias batrachus have a modified branchial cavity diverticulum for aerial respiration.

Opercular cavity

- In *Boleopthalmus* spp., the opercular chamber serves as an air breathing organ. The membrane lining the cavity is highly vascular, facilitating gas exchange.

Air Breathing Organ	Family	Species
Skin	Anguilidae, Amphipnidae	*Anguilla* sp., *Amphipnous cuchia*
Gill	Bagridae, Mastacembelidae	*Rita rita, Mastacembelus* spp.
Alimentary Canal	a. Stomach b. Intestine	*Loricaridae, Cobitidae*
Buccopharynx	Channidae, Electrophoridae	*Channa* spp., *Monopterus* spp.
Diverticulum of Branchial Cavity	Heteropneustidae, Claridae	*Heteropneustes fossilis, Clarias batrachus*
Opercular Cavity	Gobidae	*Boleopthalmus* spp.
Air Bladder	Notopteridae	*Notopterus* spp.
Neomorphic Organs	a. Arborescent Organ b. Labyrinthine Organ	*Claridae, Anabantidae*

Chapter-9

Nervous System

Introduction

- The nervous system of fishes serves as the primary mechanism for coordinating body activities and integrating responses to environmental stimuli. It comprises the Central Nervous System (CNS) and Peripheral Nervous System (PNS), each with distinct functions and structures.

Central Nervous System (CNS)

- The CNS includes the brain and spinal cord.
- It serves as the primary integrating mechanism for sensory information and motor responses.

Peripheral Nervous System (PNS)

- Consists of nerves, ganglia, receptors, and sense organs.
- Divisible into somatic and visceral nervous systems.
- Transmits sensory information from special receptor organs to the CNS and carries motor responses from the CNS to various body organs.

Autonomic Nervous System

- Regulates the activities of glands and organs.
- Comprises sympathetic and parasympathetic parts.
- Coordinates involuntary bodily functions.

Neural Components

Nerve Cells: Responsible for receiving and transmitting impulses.

Supporting Cells: Provide structural and functional support to nerve cells.

Key Processes

Impulse Reception: Nerve cells receive and transmit information.

Synapse: Transmission of information between neurons.

Membrane Potential: Difference in potential between inside and outside of nerve cells, critical for impulse transmission.

Supporting Cells

CNS: Neuroglia or simple glia.

PNS: Schwann cells and Satellite cells of ganglia.

Neuron: The Fundamental Unit of the Nervous System

- A neuron is the fundamental and functional unit of the nervous system, responsible for transmitting electrical impulses and facilitating communication between various parts of the body.

Characteristic Features of Neuron

- Neurons conduct impulses and do not divide.
- Cytoplasm contains Nissl bodies, involved in protein synthesis.

Components of Neuron

Cell Body (Perikaryon):

- Contains organelles such as mitochondria, ribosomes, golgi bodies, and lysosomes.
- Undergoes changes during nerve degeneration.

Dendrites:

- Branched structures that receive signals from other neurons.
- Site of synaptic contact.

Axon:

- Extends from the cell body and transmits electrical impulses.
- Connects with other neurons at synapses.
- Terminates in synaptic contacts known as end bulbs or terminal boutons.

Axon Transport:

- Process responsible for renewal and transport of enzymes and molecules within the neuron.

Axon Hillock and Synapse:

- **Axon hillock:** Raised cone-shaped structure where the axon connects with the soma (cell body), serving as the discharge region.
- **Synapse:** Site of communication between neurons, consisting of:
- **Presynaptic knob (terminal bouton):** Contains synaptic vesicles filled with neurotransmitters.
- **Synaptic cleft:** Narrow space between presynaptic knob and postsynaptic membrane.
- **Postsynaptic membrane:** Receptor membrane of the next neuron.

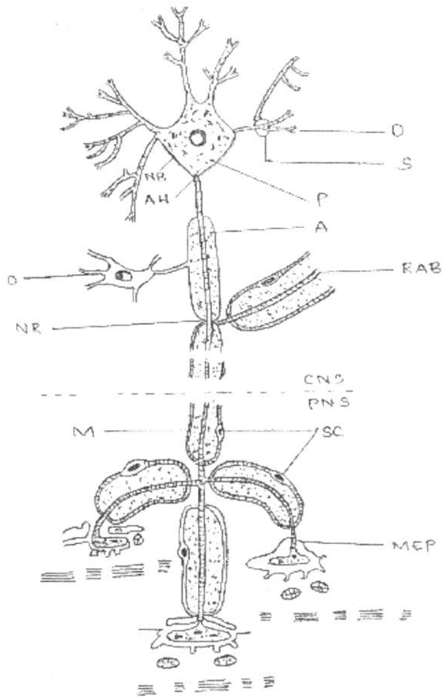

A – Axon	**AH** – Axon Hillock
CNS – Central Nervous System	**D** – Dentrites
M – Myelin	**MEP** – Motor end plante
NB – Nissl Bodies	**P** – Perikaryon
PNS – Perpheral Nervous System	**S** – Synapse
SC – Schwann cells	

Neurotransmitters

- Chemical substances stored in synaptic vesicles within the presynaptic knob.
- Include neurotransmitters such as Acetylcholine (Ach), Norepinephrine (NE), Dopamine (DA), Gamma-Aminobutyric acid (GABA), Serotonin, Glutamic acid, and Glycine.

Types of Nerves

- Sensory nerve – Receive message from sensory organs like eye, lateral line canal, barbells and sometimes taste buds.
- Motor nerve – bring message from brain to organ connected with nerves
- Intermitted nerve – lies between above two nerves

Central Nervous System

CNS consists of Brain and Spinal Cord.

A. Brain:

- During development brain distinguishes as anterior enlargement of the spinal cord.
- Brain – regulating centre for all the receptors; controls all the systems
- Brain – enclosed in a cranium
- Fish brain – small compared to body size; it does not occupy the cranial cavity completely; small gap filled with gelatinous matrix
- Brain – soft & white; covered by extensive network of blood vessels called choroid plexi

Embryonic parts		Definite parts
Prosecncephalon	Telencephalon	Forebrain
	Diencephalon	Tween brain
Mesencephalon	Mesencephalon	Midbrain
Rhombencephalon	Metencephalon	Cerebellum
	Mylencephalon	Medulla oblongata

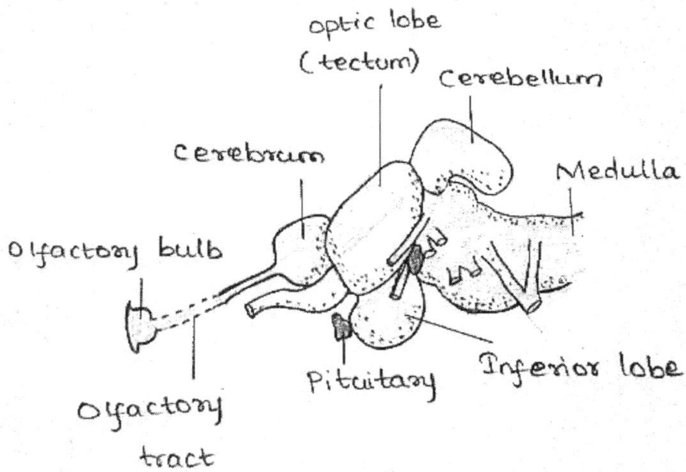

Prosencephalon or Forebrain

(a) Telencephalon

- Most anterior part of brain; responsible for reception & conduction of smell impulses
- Consists of two parts; anterior - Olfactory bulbs; posterior - 2 large cerebral Hemispheres
- Olfactory lobes are anteriorly projected and slender
- Olfactory lobes are composed of mass of nerve cells, transmitting smell impulses to other centers
- There are considerable variation in the structure of Olfactory part
- Elasmobranch and some bony fishes depending on smell for searching food have enlarged olfactory lobes
- Some fishes possess both olfactory lobes and olfactory bulbs (*Puntius ticto*) only Olfactory bulbs in *Tor tor* and only olfactory lobes in *Channa striatus*

Cerebral Hemisphere

- 2 in numbers; solid mass like, jointed on midline
- Anterior most large bundle of fibres that connects the 2 hemispheres - anterior commissure; they also connected with the diencephalon by other tracts
- Cerebral hemisphere are covered by a non-nervous membranous sheath called Pallium which is dorsally situated and subpallium is basomedially situated
- Forebrain of Actinopterygians has a single, median ventral ventricle in contrast to the paired ventral (basal) ventricles in lower fishes.
- The cerebral Hemispheres and cerebellum are most prominent in sharks and relatives and bony fishes than in the lampreys & Hagfishes. The midbrain prominent in the cyclostomes is also prominent in the Chondrichthyes. However, that of the higher bony fishes is often very large. The cavities of the brain are continuous with that of the spinal cord.

Function of Telencephalon

Olfaction; regulating aggression; sexual activity; parental behaviour; maintain body equilibrium to facilitate successful breeding; facilitate activities of lower brain center and mechanism

(b) Diencephalon

- Covered dorsally by the posterior bulgings of cerebral hemisphere
- Visible between Hemispheres and Optic lobes
- 3 parts - Dorsal Epithalamus; Lateral Thalamus & Ventral Hypothalamus

Epithalamus

- Contain choroid plexus in its roof
- Posterior part of roof elevates to form pineal gland or Epiphysis
- Contain two ganglionic mass called 'Habenulae'

Functions of Pineal body: Sensory role; secretory function

Thalamus

- Acts as a relay center for transmitting olfactory and straight body impulses to thalamomendular and thalamospinal tract
- Ventral part of thalamus bears geniculate nuclei or ganglia

Hypothalamus

- Largest & most important part of Diencephalon
- Anterior part of Hypothalamus contains preoptic area
- Ventral part of hypothalamus projects a pouch like outgrowth, Known as infundibulum
- Tip of infundibulum bears Hypophysis or Pituitary gland
- 2 important centers in Hypothalamus – nucleus preopticus & nucleus lateralis tuberis – both made of nerve cells of secretory nature
- In Osteichthyes, the lower part of the Hypothalamus extends laterally known as inferior lobe
- Sacculus venous (SV) is a highly vascular protrusion of the ventral wall of the Diencephalon
- SV – detecting changes in water pressure / secretory role

Functions of Diencephalon

Important correlation center for incoming and outgoing message concerned with internal homeostasis

Hypothalamus plays an important role in endocrine system

Mesencephalon or Midbrain

- Relatively large
- Made up of Optic tectum & Ventral tegumentum
- Fishes that feed by sight possess large optic lobes
- Optic tectum appears as two optic lobes from the dorsal side
- Optic tectum – composed of six layers
- Because of this multilayered arrangement – Optic tectum homologus with cerebral cortex of mammals

Function of Mesencephalon

- Helps in integration between the sense of equilibrium sense of vision
- Eye body coordinating center
- Important correlation centre for various types of impulses of the brain

Rhombencephalon or Hindbrain

Consists of Cerebellum or Metencephalon & Medulla Oblongata or Myelencephalon

(a) Cerebellum

- Develops as prominent dorsal outgrowths from the medulla
- Anterior part projects forward into the cavity of Optic lobes, Known as valvula cerebelli
- Elasmobranchs – size of cerebellum larger than cyclostomes
- Electric current discharging fishes such as mormyrids and electrical catfish contains well developed cerebellum dealing with electrical impulse
- Cavity – Metacoel – prominent in sharks & rays

Functions

- To control swimming equilibration
- Maintenance & coordination of muscular tonus
- Orientation in space

(b) Medulla Oblongata or Myelencephalon

- Posterior most part of brain
- Medulla is divisible into columns of nerve fibres based on the types of information transmitted
- Thus, there are visceral and somatic sensory and visceral and somatic motor coloumns
- Medulla contains nuclei of cranial nerves from II to X arranged anterio-posteriorly

Function

- Gustatory
- Sensory and sensory coordinating centres & a relying area between the remaining brain & spinal cord
- Bony fishes – respiratory; paling of body colour & Osmoregulatory

B. Spinal Cord

- Basic structure of spinal cord of fishes resembles higher vertebrates
- Runs dorsally and lengthwise of the fish body in the neural canal of the vertebral column

- In transverse section of spinal cord, two regions of clearly distinct i.e. central & peripheral
- Central region – composed of numerous nerve cells & hence looks grey & called grey matter
- Peripheral region – surrounding grey matter contains large number of myelinated nerve fibres – white matter
- Centre of grey matter – occupied by a central canal, which is filled with cerebrospinal fluid, secreted by brain
- Fish SC possess two types of cells in the grey matter- one kind of cells found in the dorsal region – controlling the movements of trunk musculature; cells found in the ventral region-fin.

Peripheral Nervous System

- It consists of cranial nerves, spinal nerves
- The bundle of nerve fibres originating from brain and spinal cord is called cranial and spinal nerves respectively.

Cranial Nerves

There are ten pairs of cranial nerves

S.No	NERVES	ORIGIN	FUNCTION
1	Olfactory nerve	Olfactory bulb particularly from lamina terminalis	Innervates snout; carries smell impulse to brain
2	Optic nerve	Ventral side of optic tectum of midbrain	Innervates retina of eye; carries visual impulses
3	Oculomotor	Optic tectum of midbrain	Innervates 4 of the 6 striated eye muscles and muscles inside eye
4	Trochlear	Dorsolateral side of brain stem under cerebellum	Innervates one of the 6 striated muscles of the eye i.e. superior oblique muscle of eye ball
5	Trigiminal	Lateral sides of the medulla oblongata; divides into ophthalmic , maxillary & mandibular branch	Sensory & motor in function; innervates anterior part of head, upper & lower jaw regions
6	Abduscence	Ventral side of MO & runs behind the Trigiminal	Innervates 6thstriated muscle i.e. Posterior rectus muscle which moves the eye ball
7	Facial	Sides of MO; is truly mixed nerve; divides into 3 branches – Supra orbital, infra orbital & hyomandibular	Innervates lateral line system & lower jaw; taste bud

S.No	NERVES	ORIGIN	FUNCTION
8	Acoustic	Sides of Medulla; runs behind the facial; 2 branches; Facial and acoustic arise from Medulla, so intimately so they are often considered as acoustic facial nerves	Vestibular branch - goes to utricular & ampullae of internal ear Saccular branch - goes to sacculus & lagena
9	Glossopharyngeal	Ventro - lateral side of MO, just behind nerve; often fused with Vagus	Innervates ventral pharyngeal mucosa and muscles of first gill slit
10	Vagus	Medulla just after glossopharyngeal; mixed nerve; divides into 5 branches	a. Supratemporal - innervates skin & lateral line b. Dorsal Recurrent - taste bud c. Body lateral line - lateral line d. Visceral - Visceral organ e. Branchial - posterior

Spinal Nerves

Paired spinal nerves arise segmentally from the cord behind the brain and carry sensory and motor fibres from and to the mass of the fish body. There are enlarged plexi at the levels of paired fins. Segmentally along the spinal cord are the relay centres (ganglia) of the spinal nerves.

Each spinal nerve contains 2 roots

- ☆ Posterior sensory root with its ganglion - composed of sensory fibres from nerve cells in the spinal nerve ganglion and connect it with the CNS
- ☆ Anterior motor root - Motor fibres passing through the anterior root have their cell bodies in the grey matter of the spinal cord.

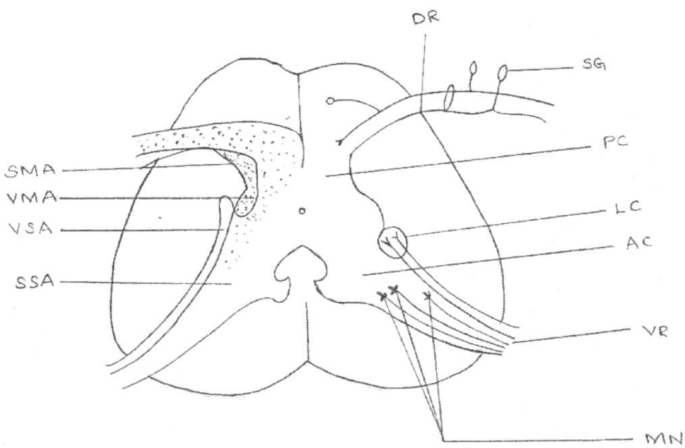

AC – anterior column, DR – dorsal root, LC – lateral column, MN – motor neuron, PC – posterior column, SG – spinal ganglion, SMA – somatic motor area, SSA – somatic sensory area, VR – ventral root, VSA – visceral sensory area.

Autonomic Nervous System

- ANS in fishes control the aperture of Iris, blood pressure, blood flow through gills for oxygenation, blood supply to various parts of the body autonomically
- It controls heart performance; gastric motility; control the function of swimbladder; colour changes; release of catecholamine from the chromaffin tissue
- ANS is the portion of PNS which conduct impulses to visceral organs, glands and blood vessels or in other words smooth muscles, cardiac muscles and glandular epithelium
- 2 types – Sympathetic and Parasympathetic.

Chapter-10

Endocrine System

Introduction

- The endocrine system is the collection of glands that produce hormones that regulate metabolism, growth and development, tissue function, sexual function, reproduction, sleep, and mood, among other things.
- The glands that secrete their products into the bloodstream and body tissues along with the central nervous system to control and regulate many kinds of body functions are known as endocrine gland.
- In fishes various endocrine gland has been found associated with different tasks and functions.

Important Endocrine Glands Found in Fishes

- Pituitary gland.
- Thyroid gland.
- Chromaffin tissue
- Interrenal tissue (Adrenocortical tissue).
- Gonads.
- Ultimobranchial tissue.
- Intestinal mucosa.

Main Tissue with Possible Endocrine Function

- Carpuscles of stannius.
- Pineal organ.
- Thymus

Endocrine Gland	Location	Origin	Secretion	Function
Pituitary Gland	Ventral surface of brain below optic chiasma	**Neurohypophysis:** floor of embryonic diencephalon	Hormones	Regulation of various endocrine functions and body homeostasis
		Adenohypophysis: dorsal evagination of buccal cavity (Rathke's pouch)		
Thyroid Gland	Near the point where afferent branchial arteries leave ventral aorta	Median evagination from the floor of the pharynx	Thyroxine	Regulation of metabolism, osmoregulation, growth, nitrogen metabolism, and maturation
Adrenal Gland	Along cardinal veins in the region of the head kidney	Three layers of cells along cardinal veins	Adrenaline, Cortisol	Regulation of fat, carbohydrate, water, and protein metabolism
Islets of Langerhans	Concentrated within the pancreas of most vertebrates	Developed from tubules of digestive gland	Hormones	Regulation of blood glucose levels and other metabolic functions
Thymus	Medial wall of the branchial cavity in actinopterygian fishes	Arises from contributions of branchial pouches	Thymic hormones	Regulation of immune system and T-cell development
Ultimobranchial bodies	Between ventral wall of esophagus and sinus venosus	Develops from epithelium of ultimate gill pouch	Calcitonin	Regulation of calcium levels in blood
Intestinal mucosa	Anterior portion of small intestine and other tissue with possible endocrine function	Produces hormones such as secretin and pancreozymin	Hormones	Regulation of pancreatic secretions through interaction with nervous controls
Corpuscles of Stannius	Attached to or embedded in kidneys of holosteans and teleosts	Outgrowths from pronephric or mesonephric duct of the kidney	Hypocalcin	Regulation of calcium balance
Pineal Gland	Posteriormid dorsal evagination of the epithalamus	Dorsal evagination of the epithalamus	Melatonin	Photosensory and secretory function, possible role in controlling reproduction and migration
Urohypophysis	Swelling at posterior end of caudal spinal cord (tail)	Posterior end of the caudal spinal cord	Urotensins	Metabolic regulations

Chapter-11

Skeletal System

The skeletal system of fish plays a crucial role in providing support, protection, and facilitating movement. It is composed of various structures including bone, cartilage, connective tissues, and specialized components like scales and teeth.

Components of the Skeletal System

Notochord

- The notochord is a flexible rod-like structure present in the early embryo of all chordates, including fish.
- It provides support and serves as a precursor to the vertebral column in vertebrates.

Bone and Cartilage

- Fish possess both bone and cartilage in their skeletal system.
- Bones provide structural support and protection to vital organs.
- Cartilage provides flexibility and support, especially in areas where movement is required.

Non-Bony Scale

- Fish have scales covering their bodies, which provide protection against predators and environmental factors.
- These scales are composed of non-bony materials and vary in size and shape among different species.

Teeth

- Fish have specialized structures in their mouths for feeding, including teeth.
- Teeth may vary in shape and function depending on the species and diet of the fish.
- Components such as enamel and dentin contribute to the hardness and durability of fish teeth.

Fin Rays

- Fins are important for locomotion and maneuverability in water.
- Fin rays provide support and structure to the fins, allowing fish to control their movements effectively.

The Skeleton of Vertebrates is broadly divided into two parts:

(i) Axial Skeleton

- Consists of the skull, vertebrae, and ribs.
- The skull protects the brain and sensory organs and provides attachment sites for muscles.
- Vertebrae provide support and flexibility to the body and protect the spinal cord.

Ribs protect internal organs and assist in respiration.

(ii) Appendicular Skeleton

- Includes the pectoral and pelvic girdles and bones of the appendages (fins).
- The pectoral and pelvic girdles anchor the fins to the body and facilitate movement.
- Bones of the appendages provide support and control for fin movement.

Main types	Types	
Axial skeleton	1.Skull 2.Vertebral column 3.Ribs & Intermuscular bones	a)Neurocranium – Ethmoid, Orbital, Otic, Basicranial b)Branchiocranium – mandibular arch, Teeth, Palatine and Hyoid arch, Opercular and branchial series a) Precaudal vertebrae b) Postcaudal vertebrae
Appendicular skeleton	1.Pectoral 2.Pelvic 3.Dorsal and Anal fins	Pectoral Gridle Pelvic Gridle

A Generalised Bony Fish Skeleton

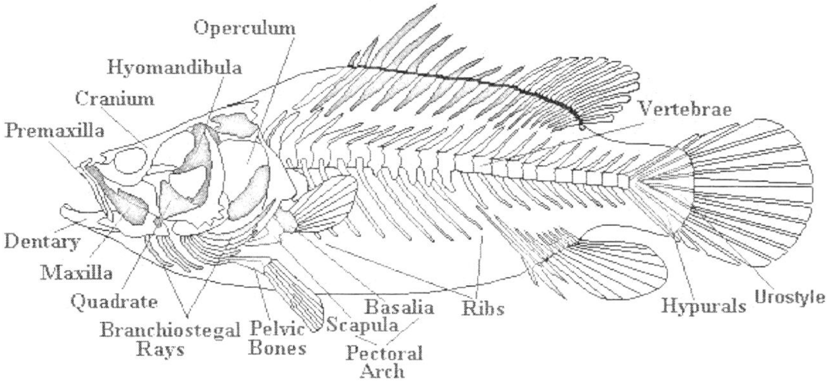

Axial Firm Skeleton

- Includes Skull, Vertebral column, ribs and intermuscular bones.

1. Skull

Bony fish

- Skull is composed of two distinctive parts.
  - ☆ Neurocranium
  - ☆ Brachiocranium

(i) Neurocranium : (Two Major Parts)

A series of inner (endosteal) bony elements that provide a floor to the brain case and surround and protect the olfactory, optic and otic capsules and anterior parts of the notochord.

A series of outer (ectosteal) dermal bone that roof the braincase, and give form to the face.

Bones of Neurocranium is Grouped by Location into the Following Regions:

- Olfactory (Nasal area)
- Orbital (about the eye)
- Otic (about the ear)
- Basicranial

(ii) Branchiocranium : (Three Regions)

- Jaw or mandibular
- Hyal (the jaw – supporting hyoid arch and the bones of the gill covering opercular series)
- Branchial (the gill arches)

Skull of Shark

- Skull is composed of a single unit cartilaginous cranium (chondrocranium) and the visceral arches (branchiocranium) and their derivatives.
- The chondrocranium has an incompletely roofed brain, ears and eyes.
- Because of the large opening that renders the neurocranial roof in these fishes incomplete, it is said that they have holes in their heads.
- Visceral arches are thought to have numbered eight primitively.
- The first become lost or is represented by the labial cartilages of those sharks that have them.
- The 2nd arch is modified into the upper jaw and the lower jaw.
- The 3rd arch in this hypothetical series is the hyoid, which spends the jaws and supports the tongue.
- The 4th through the eighth visual arches are the first through the 5th gill aches, respectively.

2. Vertebral Column & Ribs

- The backbone of a fish is composed of a series of segments, the vertebrae.
- There is one vertebra per body segment.
- Throughout the length of the trunk, the bodies of the vertebrae (vertebral centra) often have lateral processes that bear ribs.
- Throughout the length of the column, the vertebrae also form, above the centra, a series of arches that protect the spinal cord.
- Below each cranium in the tail there is an arch that partly encases main axial blood vessels.
- The rays of the caudal are supported by altered vertebral elements.

Bony Fishes

- Ribs of fishes according to the location, are of two types. (Dorsal & ventral)
- Ventral ribs are reportedly limited to bony fishes (Osteichthyes) are brone by the parapophyses of the trunk vertebrae. They develop within the connective tissue position of the lateral muscle bundles at the peritoneum.
- Dorsal ribs are also in the trunk region but lie in the horizontal skeletogenous septrum at junctions of this septum with the myosepta; hence they are often called intermuscular bones.

3. Intermuscular Bones

- Many bony fishes have small, splint bones of assorted shapes in the myosepta

  Eg : Herrings, Carps, Salmons
- Trout – bones straight, running from lateral and caudal from the vertebral column with which each intermuscular bones has a ligmental connection.

- Herrings – C Shaped
- Suckers – Forked or Y – Shaped

Appendicular Skeleton

Skeletal supports of the median and paired fins differs fundamentally in that the pectoral and pelvic fins are supported by gridles whereas the unpaired fins are not.

Pectoral and Pelvic gridles are Absent in Agnathans.

Dorsal and Anal Fin supports:

- Dorsal and Anal fin supports of bony fishes are a three bone series
- Inwardly, lying in the median skeletogenus septum between two adjacent vertebral spines, is a proximal pterygiophore (axonost) derived from an interneural dorsally, or an interhemal ventrally.
- Intemediately, enroute to the fin ray to be supported, is a pterygiophore
- Outermost in the series, articulating with the fin ray is a distal pterygiophore

Caudal Fin Supports

Major types of Caudal Fin

(i) Protocercal

- Fin is primitively symmetrical
- Fish that have notochords, not true vertebral column have protocercal tail
- Caudal fins divides equally into dorsal and ventral points.

  Eg: Lamprey and Hagfishes

(ii) Homocercal

- Exterior of fin is symmetrical
- The vertebral column ends at the base of the fin without extending into it.
- Both upper and lower lobe of caudal fin are in equal length

  Eg: Bony fishes

(iii) Heterocercal

- Fin is asymmetrical. Longer fin can be the upper dorsal lobe (epicercal) or the lower ventral lobe (hypocercal)

  Eg: Epicercal – Sharks and stutgeons

  Hypocercal – Some flying fishes

(iv) Diphycercal

- Vertebrae extend into fin, ends in abrupt point.

  Eg: Coelacanth

(v) Isocercal

- Vertebrae extend into fin, symmetrical internally and externally, regardless of shape of fin.

  Eg: Atlantic cod

(vi) Leptocercal

- Fin is long and tapering, Whiplike appearance

  Eg: Ratfish

(vii) Gephyrocercal

- No actual tail fin
- Posterior end of the vertebral column lost including the hypural plates.
- Dorsal and anal fins have grown around the posterior end of the fish to form this type of tail.

  Eg: Sunfish

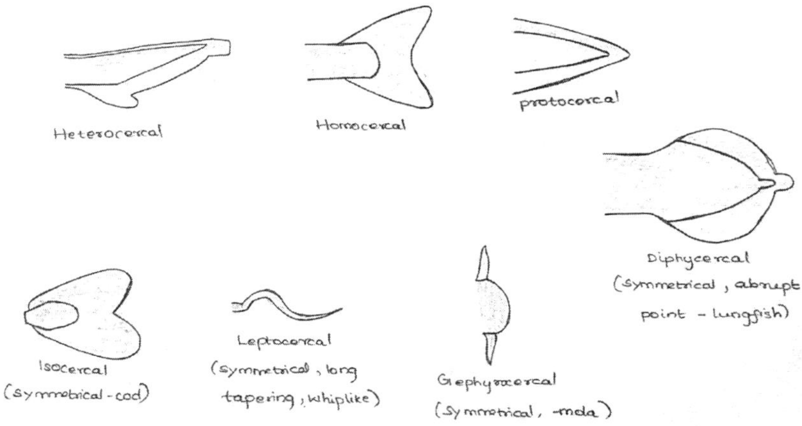

Pectoral Fin Supports (Pectoral Gridle)

- Pectoral fin in bony fishes does not have attachment to the vertebral column.
- It attaches to the back of the skull by the post temporal bone.
- The elements present in most bony fishes as cartilage – bones are the paired carocoid and scapulae and 4 pairs of radials and as paired dermal bones, the post temporals, Super cleithra, Cleithra, and Post cleithra
- The pectoral gridle is connected to the rays of the pectoral fin by various intermediate skeletal elements.
- In Sharks and relatives these are in outward sequence on each side of the fishes, a propterygum, mesopterygium and metapterygium and numerous radials which in turn bear the fin rays.

Pelvic Fin Gridle

- The pelvic gridle is usually not attached to the vertebral column
- In Sharks, the pelvic gridle consists of the isochiopubic cartilage, which floats freely in the muscle of the posterior region of the body
- In bony fishes there is a separate pelvic bone, a basipterygium and radials to which the pelvic fin ray attaches
- In the higher bony fishes, the pelvic fin radials disappear and the fin rays articulate with the bastipterygia
- Normally, pelvic fin is located in thoracic region
- Some fishes lack pelvic fins (eels, needle fish, puffer fish)

Chapter-12

Sensory Organs of Fish

Sense organs are highly specialized organs that receive physical or chemical stimuli from the environment.

- Most obvious sense organs are eye, nose, ear and skin which receive stimuli for vision, affection, audition and touch respectively.
- Less obvious sense organs are lateral line, sensory crypts, Ampullae of lorenzini in sharks and rays.

1. Eye (Photoreceptor)

Eye Structure in Fish

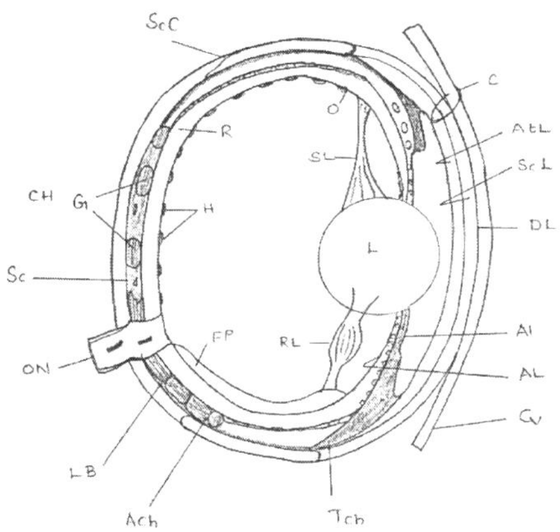

Component	Description
Cornea	Anterior most part of eye Mostly transparent, without pigment in Lamprey, shark, and most bony fishes
Sclerotic Coat	Tough and highly vascularized layer surrounding the eyeball - Supported by fibrous tissues in Elasmobranch
Choroid Layer	Present below the sclerotic coat, richly vascularized Contains a choroid gland providing oxygen to the retina
Iris	Forms the pupil and regulates the amount of light reaching the retina Iris lacks muscles in Cyclostomes
Lens	Firm, transparent ball made up of non-collagenous protein Bony Fishes have roughly spherical lens Horizontally compressed in sharks and rays
Retina	Most important and sensitive part of the eye Composed of several layers containing specialized nerve cells

Specialized Nerve Cells in Retina

Nerve Cell Type	Description
Visual Cells	Two types: Rod and Cone cells Rod cells detect light intensity, while cone cells distinguish color
Horizontal Cells	Interconnecting neurons in the retina
Bipolar Cells	Neurons that transmit signals from the photoreceptor cells to ganglion cells
Amacrine Cells	Modulate the signals sent from the bipolar cells to the ganglion cells

Light-Sensitive Pigments in Fish Retina

Pigment	Description
Rhodopsin	Purple-colored pigment found in the retina of marine Actinopterygii (bony fishes)
Porphyrospsin	Rose-colored pigment found in the retina of freshwater Actinopterygii (bony fishes)
Function	Involved in vision through photochemical processes Chemical changes are transformed into electrical impulses

Vision Process

- Involves reactions in the light-sensitive pigments of rods and cones
- Chemical changes are transformed into electrical impulses in the retina, which are transmitted via the optic nerve to the brain.

2. Olfactory Organs (Chemoreceptor)

The organ of smell in fish is a pouch that opens to the water through incurrent & excurrent channels i.e. a naris or nares divided by a flap of skin. The pouch is a blind sac in all fishes except a few of the living bony fishes and the lungfishes in which the external nostril leads through the pouch to the oral cavity. The lining of the sac is typically a lamellar sensory epithelium, connected by the olfactory nerve with the forebrain. In teleost, the olfactory chamber is occupied with the olfactory chamber is occupied with the olfactory rosette. The cilia of olfactory pit drive the water & odours are perceived when the dissolved chemicals contact with olfactory Rosette. Olfactory stimuli are conveyed to the olfactory lobe of brain via the 1st cranial nerve, Olfactory nerve.

3. Inner Ear (Mechanoreceptor)

Overview

- The inner ear of fish serves as a mechanoreceptor, responsible for hearing and balance (equilibrium).
- It is a delicate membranous labyrinth filled with fluid and encapsulated in the posterior angle of the cranium.

Composition

- The inner ear comprises the upper portion, pars superior, and lower portion, pars inferior.
- Pars superior includes three semicircular canals, forming ampulla and sac-like vesicle, the utriculus.
- Pars inferior consists of two vesicles: the anterior sacculus and posterior lagena.
- The sacculus is connected to the utriculus and filled with lymph.

Otoliths in Bony Fishes

- In bony fishes, the utriculus, sacculus, and lagena comprise lapillas, sagitta, and astericus, respectively.
- These structures are calcified and secreted by ectoderm, composed of calcium carbonate, keratin, and mucopolysaccharides.
- They are called otoliths and used for age determination.

Sensory Receptors

- Ampulla contains receptor tissue with sensory cells.
- Sensory units are supplied by the VII cranial nerve (acoustic).

Functions

- The inner ear functions to detect sound and maintain balance.
- It provides a sense of direction in response to gravity when suspended in lightless, pelagic habitats.

Weberian Ossicles in Some Fishes

- Minnows, carps, catfishes, and other teleosts connect the auditory system to the swim bladder with a chain of small bones called Weberian Ossicles.

- These ossicles connect the pulsating swim bladder wall with lymph-filled canals, aiding in sound detection.

Otoliths in Sharks and Relatives

- Otoliths are small, numerous, and diffuse in sharks and their relatives.

4. Lateral Line System (Mechanoreceptor)

- It provides a distant touch sense of object
- The object may create either mechanical disturbances or their presence may be made out from the reflected waves by echolocation
- It involves sensory lines distributed on the head and body, pit organs and Ampullae of Lorenzini
- In addition to lateral line canals following sensory structures are also included in lateral line complex.
- A. Ampullae of Lorenzini are present in the head region of sharks and rays. They are small sac like structures, which open outside by small pores. These structures are filled with jelly and have several diverticulae lined with sensory epithelium. The sensory cells of epithelium are supplied by branches of facial nerve (VII). The Ampullae of Lorenzini are sensitive to mechanical and weak electrical stimuli as well as changes in salinity. They are much sensitive to difference in electric potential. They are also considered to be electro-receptors help in locating the prey.
- B. The vesicles of Savi are found in rays. They are arranged in lines and are gelatin filled follicles contain grey, granular amorphous material. The vesicles are supplied by branches of V (Trigiminal)
- C. Electro sensitive cutaneous cells are present in elephant fish and Eels
- D. The larvae of lungfishes (Dipnoi) have hair breathing sensory cells called organs of Fahrenholz resembling the lateral line organs of unknown functions
- The receptors of lateral line system are called Neuromast. Each consists of individual hair cells with an attached Cupula which is composed of mucopolysaccharides. Water currents bends the projecting cupula, which stimulate the hair cells by bending the atrtached cilia i.e. sensory hair
- The movement of cupula produced by physical stimuli from the surroundings (pressure, vibration, touch, water current) strain the sensory hairs stimulating the hair cells

Functions : to shoal/school; to find changes in the water (pressure, current etc.,) to navigate; to food; to find obstacles.

5. Cutaneous Senses (Touch & Taste)

1. Tactile organs (Mechanoreceptors)

- In some fishes, tactile sense is very well developed
- Integument is very sensitive to touch
- It is innervated by the branches of Trigiminal & Facial nerves
- Nerve endings of these branches are abundant around mouth & on the barbell, which is tactile in nature

2. Gustatory organ (Chemoreceptor)

- Taste buds are composed of 3 kinds of cells – receptor cells, supporting cells, basal cells
- In the barbell and lips of some fishes, nerve innervating taste buds give of free nerve ending to the surrounding. The taste & touch sense may thus work unitedly.

Cyprinus carpio

- It has greatly modified taste senes resembling to mammals, which can be detect sweet, salty, acid, bitter stimuli
- Taste buds are on the palatal organs in buccal cavities gill arches and rakers of cyprinids and salmonids

Bottom dwelling catfishes

- Have relatively large number of taste receptors in their skin, fin and barbells.

Chapter-13

Excretory System of Fish (Urino-Genital System)

Excretion : Biological process by which metabolic waste (especially nitrogenous waste) are removed from the body.

Excretory Organs

- **Major excretory organ :** Kidney
- **Minor excretory organ :** Gill, Skin and gut
- **Urogenital system :** In some fishes especially Elasmobranchs, there is a close association between Kidney and Genital systems.

Kidney

Main excretory organ

Kidney is followed by ureter (Mesonephric duct) and urinary Bladder

Structure of Kidney

- It is made up of individual functional or structural unit called **Nephron**
- Slender, elongate, dark red organs extending along the dorsal aspect of the body wall just ventral to the vertebrae
- Paired – Most bony fishes ; Unpaired – Coelocanth, Latimera
- In male Elasmobranch, anterior part involves in the reproductive system
- In most of fishes, anterior part of kidney has a concentration of lymphoid and Haemopoietic tissues.

Types of Kidney

1. Pronephros

Kidney is with large number of tubules or nephrons.

Tubules of anterior region develop and become functional in early life and constitute Pronephros.

2. Mesonephros

Posterior region of Pronephric Kidney develops and becomes functional in later life and form Mesonephros.

Pronephric duct becomes Mesonephric duct.

3. Opisthonephros

The tubules of the more anterior region of Mesonephros become reduced; degenerate and converted into a lymphoid organ.

Posterior tubules take up excretory function, The posterior tubules behind the Mesonephros in fish are termed as Opisthonephros.

Classification of Kidney

- ❑ Two kidneys completely fused; no distinction between head & trunk kidney – Clupeidae
- ❑ Middle & Posterior portions fused; clear distinction – Marine cat fish & Eel
- ❑ Only posterior portions fused; no clear distinction – Belonidae, Mugilidae, Scombridae, Carangidae, Pleuronectidae
- ❑ Only extreme posterior portion fused; no clear distinction – Syngnathidae
- ❑ Completely separated – Lophiidae

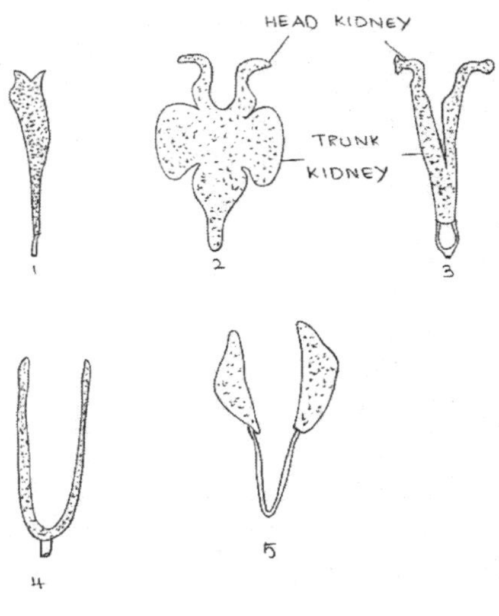

Functions of Kidney

Excrete nitrogenous waste like Ammonia

Maintain Water salt Balance (Osmoregulate)

Mesonephric Duct (Nephric Duct / Ureter)

Two Mesonephric ducts run along outer border of corresponding Kidney.

These ducts are visible in the middle & posterior region of Kidney

Two mesonephric ducts fuse to form a common duct

Urinary Bladder

Simple enlargement of common mesonephric duct or distinct sac-like structure on one side of common duct

Urinary bladder opens to exterior by a common urogenital aperture in male ; separate urinary pore in female.

Nephron

Functional or structural unit of Kidney

It consists of two parts **1) Renal Corpuscle 2) Renal tubule**

Renal Corpuscle (Bowman's Capsule) :

- Made up of two double walled Bowman's capsule & Glomerulus

Glomerulus

Mass of blood capillaries coiled within the capsule. The blood vessel that enter into the Glomerulus are Afferent Glomerulus Arteriole. The vessel that leave the Glomerulus are Efferent Glomerulus Arteriole.

- If there is no Glomerulus in Kidney – Aglomerulus Kidney Eg. Most marine fishes
- If there is Glomerulus in Kidney – glomerulus Kidney Eg. Freshwater fishes

Parts of Typical Nephron

- Renal Corpuscles
- Neck Segment
- Proximal segment
- Intermediate Segment
- Distal Segment
- Collecting duct System

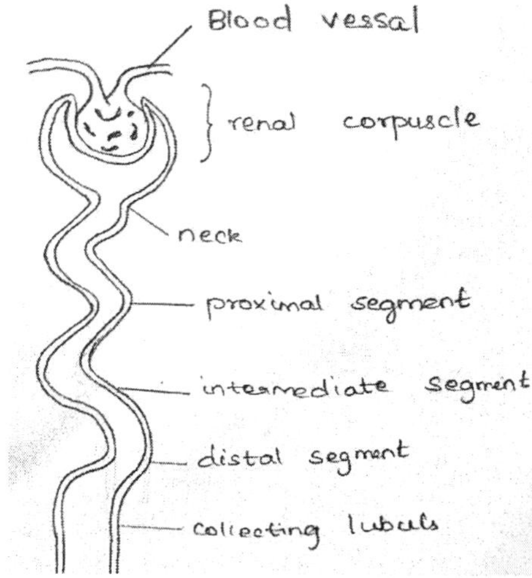

Kidney of Elasmobranch

- Opisthonephric
- Flattened; Band Shaped or strap shaped; wide posteriorly
- Anterior part of female Kidney – reduced; but in male enlarged; function as part of genital system.
- Nephrons of anterior kidney are modified to secrete seminal fluid that is received by nephric duct.
- The nephric duct is converted into sperm duct for transport of sperm and seminal fluid.
- Kidney of elasmobranch consists of a long neck, segment, initial proximal segment is composed of two cytologically distinct parts.

Chapter-14

Reproductive System

- Fishes have one pair of symmetrical gonad
- Male gonad – Testis ; Female gonad – Ovary
- Gonads are suspened from dorsal portion of the body cavity by mesentries in association with Kidney
- Testis is suspened from the dorsal portion of the body cavity by the mesentries, Mesorchium
- Ovary is suspened from the dorsal portion of the body cavity by the mesentries, Mesovarium

Reproductive System of Teleost

Male Reproductive Organ (Testis)

- 2 functions – spermatogenesis and steroidogenesis
- Comprises Testis, Sperm Duct and Seminal vesicle
- Testis in fishes – internal; longitudinal
- Originate as paired remain as such in most species
- In Gobies – testis is small and thread – like
- In Notopterus – Testis unpaired
- In Syngnathidae – Testis as simple tube
- Size and colour of testis vary according to maturity stages
- Weight – 12 % or more of the body weight
- Creamy white; Smooth

- Content – flocculent than granular
- Seminal vesicle (Sperm sac) is present only in some fishes as sperm is not stored
- From each testis Sperm duct (Vas deferens) originates
- In some fishes, Mesonephric duct unite with testis to form Vas deferens and vas efferens

Female Reproductive Organ (Ovary)

- Comprises Ovaries, oviducts
- Ovaries – internal, longitudinal, paired
- Suspended by mesenteries (Mesovarium) in the upper section of body
- Size and occupancy of the ovaries depends on sexual maturity
- May form 70 % of the body weight when fully matured
- Colour : Whitish (young); Greenish (Immature); Golden yellow (Ripe)
- Texture : Floccular, microscopic granular in juveniles to granular
- Before laying eggs enlarge and become clearer

Ovarian Types

Ovary in relation with oviduct and transport of ova, three types are there

- **Cytovarian** : ovary is connected with oviducts; both oviducts join and open through genital pore Eg. Bony fish
- **Semicytovarian** : Oocytes in place of oviduct pass through funnel groove and open into genital pore Eg. Notopteridae – Oviducts degenerates partially or completely. Ova are shed into the coelomic cavity; then carried through pores or funnel.
- **Gymnovarian** : Ovary is not connected with oviducts; ovary is hanging like a curtain; ova are shed into the coelomic cavity and carried to oviduct by cilia eg. sharks

Reproductive System of Elasmobranch

Male Reproductive Organ (Testis)

- Pair; anterior in the body cavity; suspended by Mesovarium
- Right is larger than left
- Sperms (spermatozoa) from a testis move towards exterior in the anterior part of the organ first through coiled tubules (vas efferens, modified mesonephric tubules)

Testis → Vas Efferens → Epididymis → Seminal Vesicle
↓
Urogenital Pore ← Urogenital Papilla ← Urogenital Sinus

- Vas deferens and Seminal Vesicle are clearly visible as sperms are stored
- Seminal vesicle diameter is more than sperm duct
- From cloaca, sperm enter grooves of the clasper and is transferred to female
- Associated with the clasper, under skin of pelvic fin and abdomen, a glandular sac is there for secreting lubricating fluid.

Female Reproductive Organ (Ovary)

- Paired; anterior in the body cavity; suspended by Mesovarium

Ovary → Ostium → Shell gland → Oviduct → Urogenital Sinus → Urogenital papilla → Urogenital pore

Chapter-15

Reproductive Biology of Fishes

Sexual Dimorphism

- Sexual Dimorphism is the morphological structure of each male and female fish
- Primary sexual characters (sex organs – testis & ovary) are concerned with the reproductive process
- Secondary sexual characters have no primary relationship with the reproductive process. They are accessory to spawning.

Functions of Secondary Sexual Characters

- Recognition of opposite sex
- Help in the act of copulation
- Transfer of sperm to female
- Facilitate parental care

1. Size of Fish

- Tilapia – male big ; female small
- Major carps – Female big ; Male small
- Catfish – Large male – protective adaption and parental care
- Angler fish – male dwarf and parasitic on female

2. Length / Shape / Texture of Fins

- *Labeo dero* – In male anterior part of dorsal fin has an elongated lobe

- Fantail darter – In male dorsal fin spines end in fleshy knobs
- Tibetan loach – In male pectoral fin rays are thickened like spine
- Tench – In male pelvic fin has strong spines
- Major carps – In male, pectoral fin is long and develop roughness inside the pectoral fin during the breeding season
- Sword tail – In male last fin rays of anal fin are modified into an intromittent organ, Gonopodium
- Shark – In male pelvic fin is modified into an intromittent organ, Clasper. The clasper transfers sperm from male body to the female body.

3. Colouration

- In some species (eg. Salmonids and cyprinids), male is brighter during breeding season
- Fighter fish – male is brighter than female
- In some fishes, tubercles (nuptical organ) develop on the head before spawning, after spawning they disappear.

4. Abdomen

- In crabs, Male – narrow abdomen; Female – Wide abdomen

5. Genital papilla

- In *Clarias* sp. and *Heteropneustes* sp, elongated and cylindrical papilla develop in male and button shaped genital papilla in female

6. Ovipositor

- Bitterlings (Rhodeus) – ready to spawn female has long tubes from genital aperture for laying eggs into the mantle cavity through the inhalant siphon of fresh water mussel

7. Shape of Head

- Male salmon – snout develops like hook
- In male Dolphinfish (*Coryphaena hippurus*), the forehead is very steep

Maturity Stages

Maturation : It refers to the cyclic morphological changes, which the male and female gonads undergo to attain full growth and ripeness.

Uses of Maturity Information

- To calculate length and age at first spawning
- To estimate peak spawning times and areas
- To calculate spawner biomass
- To determine stock size at capture

- To study the duration of ovary / testis cycle

Maturity Stages of the Gonods of Tropical Fishes

Stage	Female	Male
Stage 1: Immature Virgins	The ovaries appear as two pinkish translucent jelly like structure united at the posterior end. Ovaries very small, elongated and cylindrical rather oblong in shape, entire gonad occupies ¼ of the body cavity of fish. Oocytes under magnification appear irregular, transparent and with central nucleus	The testis is very small, Pinkish and translucent, 3 mm to 5 mm, in length and occupy the same portion of the body cavity.
Stage 2 Maturing Virgins	Ovaries slightly enlarged occupying more than 1/3 of the body cavity. At this stage ovarian lobe becomes cylindrical and appears dull reddish. The two lobes are unequal in size, the right lobe is larger than left and it occupies almost half of the body cavity. Under microscope, the central portion of the eggs appear darker	Testis pinkish and Opaque, still small, slightly distended but not soft.
Stage 3 Maturing Ripening	Ovaries enlarged and occupy ¾ or more of the body cavity, pinkish yellow in colour. Ovarian wall becomes thin, ova round, opaque and appear as dark bodies under microscope	Testis whitish, outer margin slightly irregular, found extending ¾ of the body cavity
Stage 4 Ripe	Ovaries very much enlarged, massive ovarian lobes, fully packed with ripe eggs, occupying the whole of the body cavity, with numerous blood vessels ramifying over their surface. Yellow in colour, eggs rounded, large and opaque some ova visible to the exterior to the vent region	Testis found extending the whole length of the body cavity very soft and cream in colour, irregular outer margin with wrinkles
Stage 5 Spent	Ovaries highly shrunken, flesh covered and collapsed occupying almost half of the body cavity. It appears, flaccid and loosely packed with primary oocytes. Blood vessels seen on surface, often some residual eggs are available in the ovarian sac. In recently spawned fish the ovaries have remnants of mature ova of a maximum diameter of 550 micron as resorbing and disintegrating opaque structure	Testis is seen extending nearly half of their body cavity, dull white and shrunken

Maturity Stages in Temperate Fishes

There are eight maturity stages for the temperate water fishes (ICES, 1962).

- Juvenile
- Developing virgin / Resting spent
- Ripening 1
- Ripening 2
- Ripe
- Running
- Spent
- Spent recovering

Gonadosomatic Index

Gonadosomatic Index : It is an index used to assess the spawning season of a fish. It is inversely related to the Gastro somatic Index. It is calculated by

Gondao somatic Index = Gonad weight / Body weight * 100

Ponderal Index

- The condition factor or Ponderal index, or coefficient of correlation expresses the condition of a fish, such as the degree of well being, relative robustness, plumpness or fatness in numerical terms.
- The condition factor used to determined from length and weight of the fish.
- Ponderal index or condition factor Where L is length in cms and W is weight of fish in grams. The cube of length is taken because the growth in weight is proportionate to the growth in volume.
- Condition factor is generally used by fish biologist as an indication of the health of a fish population.

$K = W^* 100/ L^3$ Where, W = Weight of the fish in (g), L = Total length of fish in cm.

Relative Condition Factor

It indicates the overall well-being/ health status/ fattiness of the fish.

K_n can be calculated as , $K_n = W/W^\wedge$

Where, W = observed weight of the fish

W^ = calculated weight of the fish.

If K_n is1, the well-being of the fish is at optimal level

K_n value is usually between 0.9 and 1. If it is below 0.9 the well being of the fish is not optimum.

Value of k differs with season and influenced by maturity of gonads and spawning.

It is maximum during spawning.

Length – Weight Relationship

This relationship is normally determined by the following formula

$W=a.L^b$

Where, W = weight, L = length, a = constant, b = slope

b value is usually between 2.5 – 4.0

For an ideal fish which maintains the same slope b = 3 indicates that it grows symmetrically or isometrically, provided its specific gravity remains constant. This is called **Cube Law** (Isometric growth)

- Values other than 3 indicate allometric growth.
- If the b value is above 3, weight in growth is higher compared to length in growth-positive Allometric.
- If the b value is less than 3, length in growth is ahead of weight in growth-Negative Allometric.

Fecundity

Absolute Fecundity: Number of eggs found in the ovary prior to spawning or at the time of spawning

Absolute Fecundity = Number of eggs in sample / Volume of sample * Volume of ovary

Relative Fecundity : Number of eggs found in the ovary per unit body weight.

Relative Fecundity = Absolute Fecundity / Body Weight

1. Modified Gilson's Fluid

- It was recommended that Simpson's modification of Gilson's fluid should be used in the fecundity estimation
- 60% alcohol - 100 ml
- water - 880 ml
- 80% nitric acid - 15 ml
- glacial acetic acid - 18 ml
- mercuric chloride - 20 g

Gravimetric method

- Gravimetric sampling is based on weighing and counting the eggs.
- After the eggs have been liberated from the ovarian tissues, they are thoroughly washed and spread on blotting paper to dry in air.
- Total number of eggs is then weighed and random samples of about 500 eggs are counted and weighed.
- The total number of eggs in an ovary is then calculated from the equation $F = nG/g$

(where F = Fecundity, n= number of eggs in the subsample, G=total weight of the ovary and, g = weight of the subsample in the same unit.)

Volumetric Method

- After separation in Gilson's fluid, the cleaned eggs are put in a measuring cylinder and made up to a known volume with water.
- Subsamples are drawn by shaking the container until all the eggs are evenly distributed throughout the water.
- A subsample of known volume is again drawn with a pipette, and the number of eggs in the subsample is counted.
- Then the fecundity is calculated as follows.

$$F = nV/v$$

- (Where, n = number of eggs in the subsample, V = volume which contain all the total eggs and v = volume of the subsample).

Automatic egg counter

- Fecundity can be estimated by removing the ovaries from females in stages III to V.
- Fecundity analysis is largely confined to total spawners, because it is difficult to estimate the fecundity of partial spawners.
- All the oocytes to be spawned in one spawning cycle might not have been differentiated in the early stages in the case of partial spawners.
- In later stages, some developing eggs might be the leftover of already occurred spawning.
- Adequate estimate of annual fecundity in such species, depends upon the data on the number of spawning per year, number of eggs shed at each spawning and the relation between these factors and size & age of the fish.

Types of Fish Eggs

Pelagic Eggs

- The pelagic eggs of most species are small in size, measuring about 0.7 mm to 1.5 mm in diameter.
- A few species have larger eggs between 1.6 mm and 2.6 mm in diameter.
- All pelagic eggs are transparent and are practically spherical, except for those of anchovies which are oblong (longer than broad). Occasionally eggs are found to be slightly ovoid.
- Pelagic eggs are floating type, smaller in size compared to demersal eggs. These eggs do not have adhesive membrane. They are buoyant; the buoyancy is maintained by single oil globule.
- If the oil globule is not there, high percentage of water is present which helps in floating. During floating stage, dispersion of eggs takes place. The pelagic eggs are subjected to high mortality mainly due to two factors. i.

Predation, ii. Eggs are exposed (carrying) to unfavourable conditions. But this is compensated by increased fecundity

Demersal Eggs

- The demersal eggs are generally larger than pelagic eggs which may be laid in masses or singly. These eggs are heavy or dense. Since they are heavy, they link to the bottom.
- The eggs are provided with adhesive membrane. They stick on to other objects with filamentous structure.
- Normally there is no relation between habitat and type of eggs produced. i.e. pelagic fishes can produce demersal eggs and demersal fishes can produce pelagic eggs. (Generally most common pelagic food fish have pelagic eggs)
- The pelagic sardine produces pelagic eggs, whereas herring is a pelagic fish but the eggs are demersal.
- Similarly angler fish which is a demersal fish but produce pelagic eggs. Deep sea wolf herring is a demersal fish and produces demersal eggs.

General Characteristics of Eggs

Pelagic eggs	Demersal eggs
Eggs are small in size	Eggs are bigger in size
Eggs laid singly	Eggs are laid in mass
contain small yolk content	contain more yolk content
Prolonged developmental period	Developmental period is short
Egg membrane is generally thin and smooth	Egg membrane is generally thick (with hard coverings)
Fecundity is more	Fecundity is less
No parental care	Mostly with parental care

Structure of Eggs

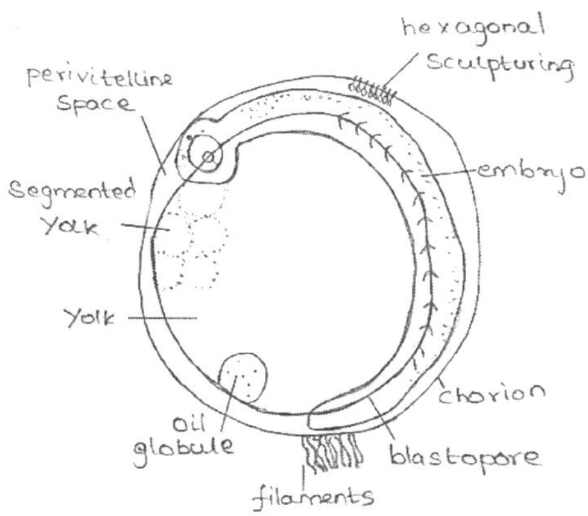

- Outer egg membrane is called Chorion
- Chorion is slightly thickened at one spot with hole, Micropyle through which sperm enters into the egg
- Within chorion, spherical yolk and surrounded by thin layer protoplasam
- Yolk may contain pigments
- Oil globule are present in buoyant eggs
- Outer membrane is smooth or sculptured or with some special structures like filaments.

Types of Eggs

(i) Based on the amount of yolk

- Alecithal – No yolk eg. Whale, Dolphin
- Microlecithal – small amount of yolk
- Macrolecithal – large amount of yolk eg. Fish egg
- Mesolecithal – Intermediate amount of yolk eg. Eggs of jawless fish, Hagfish

(ii) Based on the distribution of yolk

- Isolecithal / Homolecithal – Uniform distribution
- Telolecithal – concentrated towards vegetal pole
- Centrolecithal –Centre; surrounded by a layer of cytoplasm

(iii) Based on the presence or absence of shell

- Cleidoic – Hard shell (Calcareous)
- Non – Cleidoic – No calcareous shell

(iv) Based on the development of egg

(a) Mosaic or Determinate

Each blastomere produced by early embryonic cleavage does not have the capacity to develop into a complete embryo

(b)Regulative or Indeterminate

If a portion of egg is removed, it gives rise to seperate organism.

Larval Stages

- A larval fish, while still using its stored yolk, is called either a prolarva or a **yolk sac fry**.
- After absorption of the yolk, it is called post larva (**advanced fry**).
- Larval development continues until the fry reaches the **fingerling** stage,

Important Characters for the Identification of Fish Larvae:

Body shape

Morphometrics:

- Body depth; head size; gut length; shape of viscera; shape of viscera; fin positions including size at end of yolk sac

Meristics

- Number of myotomes / vertebrae / no. of fin rays

Pigment patterns

Specialised larval characters such as spines or opercular bones or head, shape of eyes; elongated dorsal / ventral rays or spines, extended snout.

Prolarva - Characters

- It is the just hatched larva
- It contains enormous amount of yolk. It is the food reserve for the just hatched larva. Hence, it is otherwise called "Yolk sac larva"
- Transparent with pigments in some
- Notochord and myomeres prominent with little development of cartilage or bone
- Only embryonic vertical finfold present
- Mouth and jaw not developed
- Gut as straight tube
- Blood colourless
- Circulatory / respiratory system poorly developed
- Yolk enormous – hydrostatic in function (helps in floating)
- Eye pigmentation absent
- Kidney pronephric
- < - shaped myomeres acting against notochord as skeleton

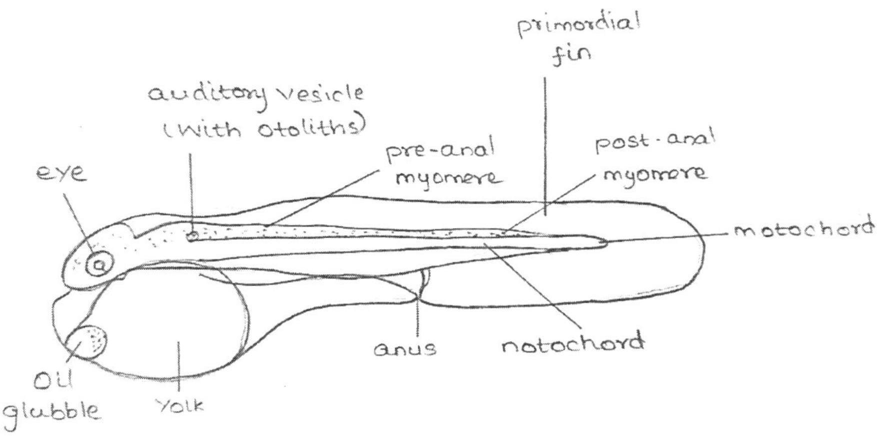

Postlarva - Characters

- Yolk resorbed (no yolk)
- Mouth formed; open
- Eyes pigmented
- Branchial arches developed
- Pectoral fins appear
- Branchial respiration replaces cutaneous respiration (using skin)
- Gill arches / Filaments appear
- Airbladder may or may not be present
- Pelvic fin or ventral fin is last one to make its appearance
- Usually fin rays are the first to appear, followed by spines and their ossification

Spawning (Fin Fish)

Spawning in finfish refers to the process by which fish release their eggs and sperm into the water to reproduce. This process is crucial for the continuation of fish species and can be influenced by a variety of environmental factors such as temperature, photoperiod (length of day), and availability of food. The spawning process is typically divided into three distinct phases: pre-spawning, spawning, and post-spawning.

1. Pre-Spawning Phase

The pre-spawning phase is the period leading up to the actual release of eggs and sperm. This phase involves several physiological and behavioral changes in fish as they prepare for reproduction.

Key Characteristics:

- **Physiological Changes**
  - ☆ **Gonad Development:** During pre-spawning, the gonads (ovaries in females and testes in males) undergo significant growth and development. This process is called gametogenesis.
  - ☆ **Hormonal Changes:** Levels of reproductive hormones such as gonadotropins, estrogen, and testosterone increase, triggering the maturation of gametes (eggs and sperm).
- **Behavioral Changes**
  - ☆ **Migration:** Many species of fish migrate to specific spawning grounds. This can involve long-distance travel from feeding grounds to spawning sites.
  - ☆ **Nesting and Territorial Behavior:** Some species prepare nests or defend territories where spawning will occur. Males often exhibit increased aggression to protect their spawning sites.

Environmental Influences:

- **Temperature:** The water temperature plays a crucial role in initiating the pre-spawning phase. Different species have specific temperature ranges that trigger reproductive behaviors.
- **Photoperiod:** The length of daylight can influence the timing of gonadal development and spawning activities.

2. Spawning Phase

The spawning phase is the period during which eggs and sperm are released into the water. This is the critical phase for fertilization and the beginning of the next generation.

Key Characteristics:

1. Release of Gametes

- **External Fertilization:** Most finfish species release their eggs and sperm into the water column where external fertilization occurs. This process is often synchronized to maximize the chances of successful fertilization.
- **Spawning Aggregations:** Many species form large groups or aggregations during spawning to increase the likelihood of egg and sperm encounters.

2. Spawning Behavior

Courtship: Fish often engage in elaborate courtship behaviors to attract mates. These behaviors can include color changes, specific movements, and sounds.

Timing: Spawning can occur at specific times of day, tides, or lunar cycles, depending on the species.

Environmental Influences:

- **Habitat:** Suitable spawning habitats are crucial for the success of this phase. These habitats can include coral reefs, riverbeds, or open water areas.
- **Water Quality:** Clean, oxygen-rich water is essential for the survival of eggs and developing larvae.

3. Post-Spawning Phase

The post-spawning phase is the period following the release of gametes. This phase involves the recovery of adult fish and the early development of offspring.

Key Characteristics:

Adult Recovery

- **Energy Replenishment:** Spawning is an energetically demanding process. After spawning, adult fish focus on feeding and restoring energy reserves.
- Return Migration: Many species migrate back to their feeding grounds after spawning.

Offspring Development:

- **Egg Incubation:** Fertilized eggs develop into larvae. The incubation period varies among species and is influenced by environmental conditions.
- **Larval Stage:** After hatching, larvae go through several stages of development before reaching juvenile and adult stages.

Environmental Influences:

- **Predation:** Eggs and larvae are vulnerable to predation. The survival rate is often low, but producing a large number of eggs compensates for this.
- **Food Availability:** The availability of suitable food sources is crucial for the growth and survival of larvae.

Spawning Stages:

- Broadly, it can be categorized into three types. They are
  - Pre-spawning
  - Spawning
  - Post spawning
- In pre-spawning, different stages of development of gonad (Testis and Ovary) takes place. During spawning stage, release of sperms and eggs occurs. Post spawning stage is the recovery stage where maturation of gonads begins from the initial stage.
- Prior to the spawning, sperms are released into the sperm duct which is called spermiation. Similarly, the eggs are released into the oviduct prior to spawning which is called ovulation.

Reproduction of Fishes

Production of new individuals or offsprings.

In this process, genetic materials are transferred from parents to the offsprings (one generation to another generation)

Age and on set of Reproduction

- Age and onset of reproduction varies between species and sex
- Environmental conditions affect the onset
- Male matures earlier than female. This is due to the lengthy physiological process for ovary and ova development and energy consuming process
- Life span of tropical fishes is short. So, they undergo early maturation and reproduction
- In temperate waters, life span of fishes (eg. *Clupeaherrangus*) is long. So, stage of maturity is very long (4 years).
- Lake trout (Salmo trutto) spawn at 5 – 7 years. Same species in river matures at 3 – 4 years

- Under favourable conditions, female undergoes somatic development (body growth) rather than gonadic maturation
- Reproduction is delayed in females, but large no. of eggs are produced
- If competition for food is low, animal gets maturity earlier.

Types of Reproduction

- Bisexual
- Hermaphroditism
- Parthenogenesis

1. Bisexual Reproduction: (Gonochorism)

- Sexes are separate

2. Hermaphroditism

- Both male and female sex organs are present in one individual animal

Bidirectional Sex Changers: where an organism has both female and male gonads, but act as either female or male during different stages in life.

(i) Synchronous Hermaphroditism

- Male and female sex organs ripe at same time.
- Usually, self-fertlization does not occur.
- Eg. Pulmonate land snails, land slugs and earthworms

(ii) Metagonous Hermaphroditism:(Asynchronous / Sequential / Consecutive)

Either male or female organ develop at a given time

When an animal starts its life as male and changes its sex into female in its later stage of life, it is called Protandrous Hermaphroditism

Eg. *Lates calcarifer* (Sea bass), *Clown fish*

When an animal starts its life as female and changes its sex into male in its later stage of life, it is called Protogynous Hermaphroditism

Eg. Sparidae, Labridae

3. Parthenogenesis

It is the development of young without fertilization

However, male is required to incite egg to develop

Resultant young ones are always female

Eg. Amazon Molly (*Poecilia formosa*)

Modes of Reproduction

- Oviparous
- Ovoviviparous
- Viviparous

1. Oviparous

- Reproduction by laying eggs. Eg. Carps, Tuna

2. Ovoviviparous

- Reproduction by giving birth to live young ones by the female
- No placental or blood connection between developing embryo and the mother
- Eg. Guppy (Poecilidae) and shark

3. Viviparous

- Reproduction by giving birth to live young ones by the female
- There is a placental or blood connection between developing embryo and the mother
- Developing embryo receives nutrients and oxygen from the mother

Four Mating Strategies

	Single Female	Multiple Females
Single male	Monogamy	Polygyny
Multiple males	Polyandry	Polygynandry

Semelparous: if its individuals spawn only once in their lifetime – Pacific salmon, squid, octopus

Iteroparous: if the individuals spawn more than once.

Reproduction Strategies

Three strategies

- Non-guarders
- Guarders
- Bearers

1. Non-guarders

78% of the fishes are non-guarders

2 types – 1) Open Substrate Spawner

2) Brood hider

Open Substrate Spawners

Lay eggs in open places – 7 types

Pelagic spawner (Pelagophils)– Buoyant eggs in water surface

- Rock and Gravel spawner with pelagic larva (Lithopelagophils) – at first eggs are adhesive; then buoyant; larva buoyant
- Rock and Gravel spawner with benthic larva (Lithophils) – Larva photophobic; hide under stones
- Non-obligatory plant spawner (Phytolithophils) – Adhesive eggs on plants; Hatchlings are photophobic and has cement gland for attachment
- Obligatory plant spawner (Phytophils) - Adhesive eggs on plants; Hatchlings are not photophobic
- Sand spawner (Psammophils) – Adhesive eggs in running water on sand or root; Hatchlings have no cement gland
- Terrestrial spawner (Aerophils) – eggs are laid outside the water; in damp soil

Brood Hiders - 5 types

- Beach spawner (*Aeropsammophils*)
- Annual fishes (*Xerophils*)
- Rock and Gravel spawner (*Lithophils*)
- Cave spawner (*Speleophils*)
- Spawning in live invertebrates

2. Guarders (22% Fishes)

Either male or female or both guard eggs 2 types

- Substrate chooser
- Nest spawner

Substrate chooser: 4 types

- Pelagic spawner
- Above water spawner
- Plant spawner
- Rock spawn

Nest spawner

- Froth nester (Bubble nest / aphrophils)
- Hole nester
- Miscellaneous material nester (polyphils)
- Rock and Gravel nester
- Anemone nester

- Plant material nester – glue making
- Plant material

3. Bearers

Carry eggs in the body during incubation 2 types –

- External bearer
- Internal bearer

External bearer

Carry eggs in the external parts of the body — 7 types

- Transformer brooder: Carry eggs in the pelvic fin or as cluster hanging from genital pore for sometime and then deposited. After depositing, they become non-guarders.
- Skin brooder
- Forehead brooder
- Gill chamber brooder
- Pouch brooder Seahorse / Pipefish – male has a brood pouch Brazilian catfish – male has enlarged lip as pouch
- Mouth brooder Tilapia and catfish
- Intestinal brooder

Internal Bearer

- **Facultative internal bearer** – actually oviparous, but accidentally becomes bearer
- **Obligate lecithotropic live bearers** – no maternal embryonic nutrient transfer
- **Viviparous trochoderms** - maternal embryonic nutrient transfer is there.

Chapter-16

Developmental Biology of Fishes

- Developmental biology of fishes – Development of fishes from Gametes to adult
- In sexual reproduction, new individuals are produced by special generative cells called Gametes.
- Male gamete – sperm; Male gonad – Testis
- Female gamate – Ova; Female gonad – Ovary

Stages in the Development of Fishes

- Gametogenesis - i) Spermatogenesis ii) Oogenesis
- Fertilization
- Cleavage
- Gastrulation
- Organogenesis
- Hatching and Larval development

I. Gametogenesis

(i) Spermatogenesis

- Spermatogenesis is the process by which spermatozoa are produced from male primordial germ cells through mitosis and meiosis.
- It is the development of the spermatozoa in the testis.
- Testis are composite organs, consisting of numerous seminiferous tubules/ seminiferous lobules converging towards common ducts–leading the mature sperm to the exterior.

- As a result of meiotic, spermatids are formed. Though spermatids possess haploid chromosomes, they do not function as male gametes. Spermatids undergo a process of differentiation to become the spermatozoa.

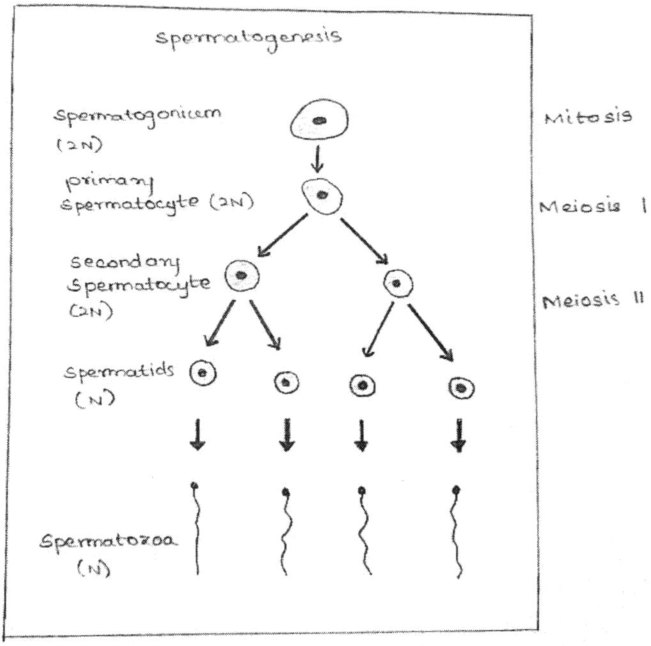

Structure of Spermatozoa (SPERM)

- Typical spermatozoa have a head, middle piece, tail/flagellum.
- Anterior tip of head is called Acrosome. It penetrates the egg to establish connection with egg cytoplasm
- Posterior part is Centriole – helps during cell division in the fertilized egg
- Tail or flagellum – longest part of sperm – its movement causes it to swim with head (Acrosome) foremost.
- Function : To fertilize egg

(ii) Oogenesis

- It is the process of production of ova in ovary.
- Oogenesis undergoes by mitotic division.
- From the primordial germ cells, Oogonia or egg mother cells are formed. They undergo mitotic division to form oocytes (Primary).
- Primary oocytes undergo 2 phases – Previtellogenesis and vitellogenesis
- During Previtellogenesis, primary oocytes increase in size / Volume
- Vitellogenesis is the process of yolk formation.

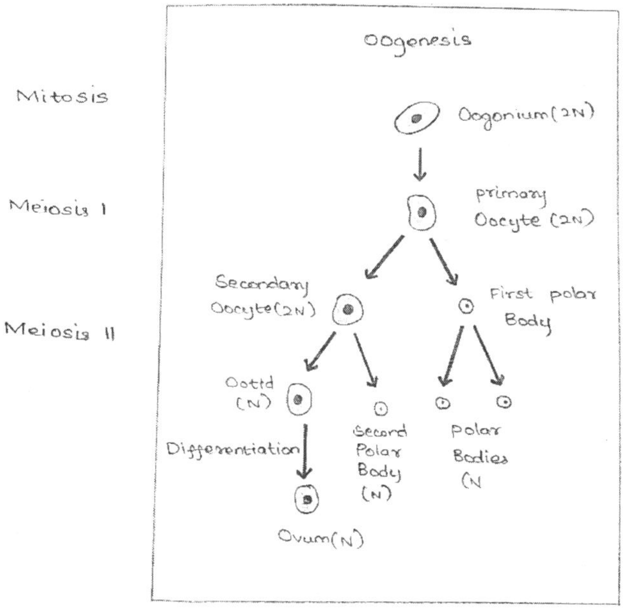

II. Fertilization

- ❑ Fertilization is the Fusion of gametes
- ❑ Consists of steps – Ovulation, spermiation and activation
- ❑ Ovulation – process of release of ovum from ovary
- ❑ Spermiation – process of release of sperm from testis.
- ❑ Fertilization is the fusion of male and female gametes to form zygote.
- ❑ Fusion activates eggs. So, egg starts to develop.
- ❑ Majority aquatic animals – External fertilization
- ❑ Sharks / Poecilids (Guppy, Molly) – Internal fertilization
- ❑ During Fertilization, sperm enters through microphyle to reach egg cytoplasm
- ❑ Fertilizin present in egg and antifertilizin in sperm reacts each other.
- ❑ Micropyle allows only one sperm to enter into egg.
- ❑ Activation is the process of initiating development in the egg; initiated / stimulated by male gamete.

III. Clevage

- ❑ Fertilization is followed by Clevage
- ❑ In embryology, cleavage is the division of cells in the early embryo. The zygotes of many species undergo rapid cell cycles with no significant growth, producing a cluster of cells the same size as the original zygote.

- The different cells derived from cleavage are called blastomeres and form a compact mass called the morula. Cleavage ends with the formation of the blastula.
- Segmentation of zygote into smaller and smaller cellular units, called Blastometres, forming an embryonic stage, Blastula.
- The blastula is a hollow sphere of cells, referred to as blastomeres, surrounding an inner fluidfilled cavity called the blastocoele

Pattern of Cleavage (2 types)

Holoblastic (total or entire cleavage) – Entire egg divides/complete cleavage of egg – Eg. Frog, Starfish

- Meroblastic (partial cleavage) – Only a portion egg divides; Incomplete cleavage. Division occurs only in animal pole. Yolk in vegetal pole is undivided. Eg. Fishes.
- The pole of the egg with the highest concentration of yolk is referred to as the vegetal pole while the opposite is referred to as the animal pole.

IV. Gastrulation

Gastrulation is a phase early in the embryonic development of most animals (including fish), during which the single-layered blastula is reorganized into a trilaminar ("three- -layered") structure known as the gastrula. These three germ layers are known as the ectoderm, mesoderm, and endoderm.

It is a dynmic process in which major presumptive organ forming area of blastula are arranged and re-organised to permit their ready conversion to the body plan of particular species.

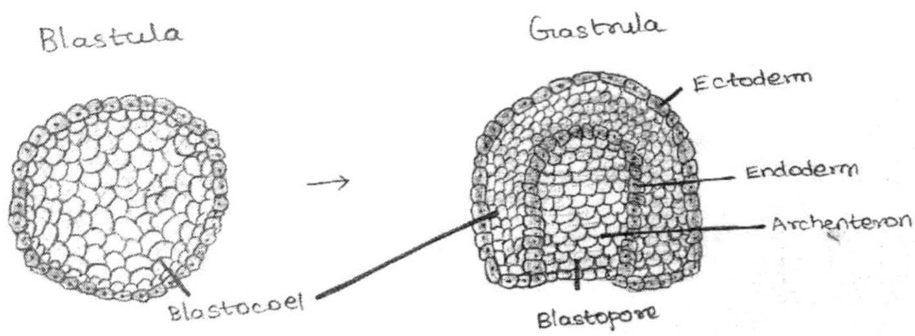

V. Organogenesis

Development of various organs from 3 germ layers of Gastrula is called Organogenesis.

Various Stages of Organogenesis

- Neurogenesis – Formation of Neural tube – further differentiation into nervous system
- Notogenesis – Formation of Notochord

- Mesogenesis – Formation of mesoderm and mesodermal organs – develops on either side of notochord between ectoderm and endoderm.
- Enterogenesis – Formation of gut from endoderm or archenterons.
- Development of eye from ectodermal derivatives
- Development of ear from ectodermal derivatives
- Development of fins and scales from ectodermal derivatives
- Heart – Lateral mesoderm
- Kidney – mesodermal
- Reproductive organs – mesodermal
- Gills – Mesoderm

VI. Hatching and Larval Development

- Hatching is a process by which animals change their life from intracapsular life to free living life process . It is the larvae come out of the egg.
- Repturing chorion or outer membrane by mechanical or enzymatic
- Enzymatic hatching – enzymes
- Mechanical hatching – pressure / tail movement

Chapter-17

Age and Growth Determination by Direct and Indirect Methods

Methods of Age Determination

Direct Method

- Rearing the fish
- Tagging or marking

Indirect Method

- Length frequency analysis
- Growth checks on hard parts

1. Direct Methods

These are the methods from which the age of the fish could be known directly.

(i) Rearing the Fish

In this method fish is reared in large sized net cages in the sea (or manageable water body like pond, reservoir). It is essential that the areas where the fishes are reared should be in natural grounds of the species and the environmental condition such as hydrological, trophic and ecological should be as in natural grounds in space and time. The length of the fish is recorded before the start of rearing and at the end point of rearing period; and the age is calculated for the rearing period.

(ii) Tagging or Marking

A direct and positive method of determining the age of fish is by the process of liberation and recovery of a targeted or marked fish of known age. This is also time consuming and costly method. This method provides an accurate base for comparing the ages determined by other methods.

2. Indirect Methods

These are the methods by which the age of fishes is determined in an indirect manner.

(i) Length Frequency Method

Collection and Processing of Length Frequency Data

- ❑ Samples for length frequency should be collected at random from the commercial catches as soon as they are landed.
- ❑ Sampling should be done before sorting into various market size groups
- ❑ Length data should be collected separately for different gears and for different mesh types of the same gear.
- ❑ Length frequency data are to be recorded twice or thrice in a week from the landing centre
- ❑ While collecting the length data, sexes should be treated separately.
- ❑ While recording the length data, total length or fork length or standard length should be taken depending on convenience.
- ❑ The data should be recorded in mm or cm in a primary register
- ❑ It is better to take length frequency data for a period of two years
- ❑ Classify the individual length recording into length groups which should not normally exceed twenty five.

The Weekly Data may be Pooled on a Monthly Basis

The data thus pooled must be drawn in the favour of frequency polygons or histogram for each month.

- ❑ **Petersen method:** Assumptions are made as to the time interval separating the various peaks of one length frequency sample, these peaks assumed to represent distinct age groups.
- ❑ **Modal progression analysis:** length frequency sample are arranged sequentially in time with this method, assumption are made as to which of the peaks can be inter-connected that belong to various samples arranged sequentially in time.
- ❑ **Integrated method of Pauly:** This method is devised by Pauly (1983) in which the above two methods are combined. The demerits of the first two

methods are set right in Pauly's integrated method in which the growth curves are drawn by curved ruler directly upon the length frequency samples sequentially arranged in time.

(ii) Growth Checks on Hard Parts

(a) Scales

The scales of the fish are embedded in the pockets of skin and the embedded part is covered with striations and concentric rings. Each scale possess a focus around which concentric rings denotes the age of the fish.

Focus: the central, small, clear area and displayed from the center of the scale

Circuli: Concentric marks around the focus

Annuli: Appears as wide circular troughs. Their number depends upon the age in years of the individual.

To study the age of fish the scales should be collected from Dorsal, Ventral, lateral and caudal regions. Before plucking the scales, the fish is first washed under tap water to remove any loose scales, slime etc. the scales are then cleaned by dipping in freshwater and rubbing between thumb and fore fingers to remove dirt and mucus. As soon as the scales are plucked from the fish, they may be washed in dilute KOH before mounting. The scale should be mounted with their convex side up on microscopic slides. Usually the scales will adhere directly to the slide once they have dried. A sample of 6 scales is generally needed.

If the scale are thick and transparent, it is necessary to treat them with caustic soda for sometime. These scales are stained with Borax carmine or Alizarine or Alizarine Sulphonate and observe rings. Scale method is suitable for Carps, Salmons, Herring, Cods etc.

No. of annuli on the scales	Estimated Age	Designation of Age group
None	Less than 1 year	0+
One	one	1
One + some circuli	More than 1 year	1+
Two	Two years old	2
Two + some circuli	Between 2 and 3	2+

(b) Otolith

Otolith are lodged in pro – otic bone. They are found in the sacculus of the inner ear. A total of three otoliths viz. Lapillus, Asteriscus and Sagitta seen in fish. Among them the largest otolith sagitta is mainly used for age determination.

Otolith bears annulations formed by the regular deposition of calcium salts. The annulations are formed when there is a differential growth rate in different season

- Transparent zone or Hyaline zone
- Opaque zone

One complete cycle of otolith growth consists in the laying down of one opaque zone and of one transparent zone. One hyaline zone and one opaque zone is taken as 1 year zone. If we get 2 opaque zone and one transparent zone the age is taken as one plus.

Preparation of Otoliths for Age Determination

Fish are first boiled in 2-5% KOH. The head bone is dissected and otoliths are removed using a pair of tweezers and preserved in paper envelops. They can also be preserved in a mixture of Ethyl Alcohol and Glycerine. To brighten the otoliths and to accentuate the appearance of growth rings, glycerine, xylol, cedar wood oil cresol, immersion oil are used. Before observing under microscope, they are first cleaned and polished. For larger otoliths, cross sections are observed through the nucleus using a knife. Otoliths can be burned over a very low flame of Bunsen burner or a spirit lamp to accentuate the appearance of the rings.

Other structures used for age determination:

- Vertebrae
- Fin rays
- Fin Spines
- Operculum

Chapter-18

Fish Migration

- Mass movement of fishes from one habitat to another with characteristic regularity in time or according to life history stage.
- Migratory behaviour of fish is a regular phenomenon. Their journey is purposed mainly for feeding and reproduction.
- In fishes various types of migratory movements are seen on a regular basis, on a particular time scale ranging from daily to annually or longer, and over a distance ranging from few meters to thousands of kilometres.

Examples of Migratory Fish Species

- Herring (*clupea harengus*)
- Salmon *(salmo* sp.*)*
- Eel *(Anguilla anguilla)*
- Hilsa (*Hilsa ilisha*)
- Three spined stickle back (*gasterosteus*)
- Lampreys (*Petromyzon marinus)*

Types of Fish Migration on the Basis of Needs

- ***Alimentary or Feeding migration**:* migration for search of feeding ground. It occurs when food resources get exhausted.
- ***Gametic or spawning migration**:* it occurs during breeding season in search for the suitable spawning ground.
- ***Climatic or seasonal migration**:* migration in search for suitable climatic condition.

- ***Osmo-regulatory migration***: migration for water and electrolytes balance from sea to fresh water and vice versa.

Types of Fish Migration Based on Spawning

Many fish migrate long distances to spawn. These migrations are classified into different categories.

According to Myers(1949) and Mc Dowall (1988) the categories are :

- **Diadromous migration**
  - ☆ Anadromous migration
  - ☆ Catadromous migration
- **Potamodromous migration**
- **Oceanodromous migration**

1. Diadromous Migration

Truly migratory fishes that travel between the sea and freshwater comes under diadromous migration.

This migration is of two types :

- Anadromous migration
- Catadromous migration

(i) Anadromous Migration

- It is the migration of marine fishes from sea to fresh water for spawning.
- Fishes spend most of their life living and feeding in sea.
- They only migrate during breeding season to the river for spawning ground.

Examples:

- Salmon
- Hilsa
- Lamprey
- American shad

Salmon migrate for breeding during winter from sea to river. While migrating, some physiological changes occurs:

- Stops feeding during journey.
- Changes colour from silver to dull reddish brown.
- Gonads mature.

They select suitable spawning ground and make a saucer-like nest in which female lays eggs

Juvenile larva hatched out from the egg known as Alevins. Alveins then transform into parr and metamorphosed into adult when return to the sea.

(ii) Catadromous Migration

- It is the migration of fresh water fishes from river to sea during breeding season for spawning.

Example:

Eels

Both European eel (*Anguilla anguilla or Anguilla vulgaris*) and the American eel (*Anguilla rostrata*) migrate from the continental rivers to Sargasso Sea off Bermuda in south Atlantic for spawning, crossing Atlantic Ocean.

Before and during migration some physiological changes occur in their bodies:

- Deposit large amount of fat in their bodies which serves as reserve food during the journey.
- Colour changes from yellow to metallic silvery grey.
- Digestive tract shrinks and stops feeding.
- Eyes get enlarged and vision sharpens. Other sensory organs also become sensitive.
- Skin serves respiratory organ.
- Gonads get matured and enlarged.
- The lay eggs in suitable spawning ground and are fertilized by males. After spawning they die. The larva hatch out and develop into young eel and finally return to river.

2. Potamodromous Migration

- Migration is limited to Freshwater only.
- It is the Freshwater migration of fishes from one habitat to another feeding or spawning.

  Eg.: Carps, Catfish

3. Oceanodromous Migration

- It is the migration of fish within sea in search of suitable feeding and spawning ground.
- Fish are born near spawning grounds, then drift on ocean currents as larvae before settling as juveniles to grow into adults before migrating back to spawning grounds.

Examples:

- Tuna
- Clupea

Vertical Migration:

- Many species of fish and invertebrates make dial vertical migration (with 24 hr periodicity) usually towards the surface at dusk and towards the bottom at a dawn.
- To feed phytoplankton at night; to avoid predators during day.
- Herring-moves up or down 100-200m during the dial cycle.
- Swordfish-moves from the surface to 600m between day night and day.

Horizontal Migration:

- Some fishes migrate horizontally to feed and to avoid predators.
- Reef fish moves on and off the reef with a 24hr periodicity, feeding by day and hiding at night.
- Larger seasonal horizontal migrations occur that are related to spawning and feeding.

Advantages of Migration:

- Animals take advantage of additional habitat and avoid adverse conditions.
- Animals are placed in the best or necessary location for given biological activities at the appropriate time.

Disadvantages of Migration:

- Increase in the expenditure of energy.
- Exposure to predators along the migration route.

Chapter-19

Marking and Tagging

Marking : Any procedure or technique that makes fish identifiable either as an individual or as a member of the batch

Tagging : Attaching foreign objects bearing a number to the body of the fish to make them identifiable

Purpose of Marking & Tagging

- To study the population patterns
- To study the migratory pattern
- Growth and age determination
- Survival and growth of hatchery plants

Marking Techniques

I. Group Marking Techniques

1. Fin Clipping

- Part of the fin is cut by using scissors.
- Most popular technique
- Regeneration is quicker among young fish
- Pelvic fin is usually used
- Damaging the tail may affect the mobility

Disadvantage:

- Suitable only for short term marking
- Fins may regenerate
- Getting infected , affecting the physiology of the fish

2. Opercular and Fin Punch

Hole is made in opercular bone or fins using punching machine

Disadvantage:

- Suitable only for short term marking
- Not suitable with fleshy opercular bones
- Mark heals fairly quickly , marking detection difficult
- Getting infected , affecting the physiology of the fish

3. Branding

- Here, hot or very cold rods are used to brand the fish. It is done below the dorsal fin or in the caudal peduncle region.
- **Hot Branding -** using heated pencil or electrically heated wire
- **Cold Branding -** using a mixture of Ethanol and Dry ice or Acetone and Dry ice; liquid Nitrogen or Solid CO_2

Disadvantage:

- Suitable only for short term marking
- Risk of fungal infection

4. Tatooing

The pigments are implanted just below the skin by using a needle or vibrator. Indian ink or Trypan Blue in Titanium is used.

Disadvantages

- Suitable only for short term marking
- Not suitable for fish with large scales
- If care not taken, marking mortality may be high

5. Subcutaneous Injection

- Subcutaneous injection or materials are inoculated just below the skin by using a syringe. Pigments used are Fast Blue and Hydrated Chromium Oxide and Alcian Blue.

6. Injection with Liquid Latex

Coloured latex is injected just below the dorsal fin. This is useful for transparent bodied animals

7. Injection with Vital Stain

This is used to study the organogenesis. Vital stains like Bismark Brown or Neutral red are given by immersion. Sudan Blue is given through feed.

8. Injection with Fluorescent Materials

Fluorescent materials-like tetracycline are injected into the body. This technique is used for invertebrates (Crustaceans). In the Tail, Cephalothorax and the Abdominal region the fluorescent materials are injected.

II. Individual Marking Techniques

Here, individual organisms are given a separate no. or code to identify them. There are two techniques.

- ❑ Internal Marking Technique
- ❑ External Marking Technique

1. Internal Marking Technique

A metal or plastic coded material will be inserted inside the body and the inserted material can be identified by X-ray method or by radioactive method or by magnetic principle.

2. External Marking Technique

They are fixed on the surface of the body in the dorsal fin area or head region or in jaw area. Various types of external tags are available in the market

(i)T – Bar Tags

These are numbered plastic tags with a 'T' shaped head. These tags are cost effective and popular on fish, crustacean species. T-bar tags are convenient even in some shellfish species where large numbers of fishes may need to be tagged in a short space of time and/or holding time is critical for fish survival.

(ii) Plastic tipped dart tags

These tags are very popular with recreational fish tagging programs for a wide variety of species due to ease of use & lower unit cost of applicators compared to T-bar tags. The large dart tags are used on finfish and sharks with a size of about 60 cm including tuna, barramundi, tarpon and king mackerel.

(iii) Stainless steel head dart tags (SSD)

These are used for marking large fishes like billfish and sharks which cannot be easily brought aboard. The tag head is specially ground and sharpened to anchor smoothly in firm tissue or muscle.

(iv) Body cavity tags

Body cavity internal anchor tags have extremely long term tag retention. Hence, they can be used for the estimate of population size or exploitation rate. The tag is designed in such way that the anchor lies parallel with the inside abdomen wall and the external marker lies backwards.

Glossary

Adipose eyelids: Transparent membranes covering partly the surface of eye

Adipose fin: Fin which is fleshy or fatty and present behind dorsal fin

Air bladder: A membraneous gas filled sac lying just beneath the back bone either open or closed

Alimentary Canal: Also known as the digestive tract, it is a tube connected with the digestion and absorption of food, extending from the mouth to the anus.

Anterio-inferior: That which is situated below anterior end of head

Anterio-transverse: That which is situated in the middle of anterior end of head

Anus: External opening of the intestine often referred to as the vent

Asymmetry: Lack of symmetry; absence of a plane dividing an organism into similar halves.

Atrium: The upper chamber of the heart that receives blood from veins.

Axon Hillock: Raised cone-shaped structure where the axon connects with the soma (cell body), serving as the discharge region.

Back: The dorsal side of a fish

Bony plates: Hard plate like structures. which are modified from scales (see carapace) which encase body

Bony rings: Hard ring - like structureswhich are modified from scale which encase body

Branched ray: A soft ray which forks into 2 or more parts distally

Branchial: Pertaining to the gills

Branchiostegals: Bony rays supporting the gill membrane

Buccal: Pertaining to the mouth cavity

Buccopharynx: The part of the alimentary canal comprising the lips, buccal cavity, and pharynx.

Bulbus Arteriosus: The portion of the fish heart that connects the ventricle to the ventral aorta, facilitating blood flow.

Capillaries: Small blood vessels that allow for exchange of nutrients and gases between blood and tissues.

Cardiform: Elongate conical teeth, Coarse and sharp small teeth

Carinate: Having a keel or ridge along the midline

Cartilaginous: Composed of cartilage

Caudal: Pertaining to the tail

Caudal fin (Shapes): Pointed, truncate, emarginated, forked, lunate, wedge shaped

Caudal peduncle: The narrow terminal part of body between the end of dorsal fin and the base of caudal fin

Cephalic: Pertaining to the head

Chondrichthyes: A class of cartilaginous fishes, including sharks, rays, and chimaeras, characterized by a skeleton composed of cartilage rather than bone.

Chorion: The outer membrane surrounding fish eggs.

Chromaffin Tissue: Tissue associated with the adrenal glands that secretes hormones such as adrenaline.

Clasper: Extensions of the pelvic fins in male sharks

Closed Circulatory System: A circulatory system in which blood circulates entirely within a series of vessels and is distinct from the interstitial fluid.

Coelomic Cavity: The body cavity of many metazoan animals, located between the intestinal canal and the body wall, filled with coelomic fluid.

Compressed: A body form flattened from side to side

Conical teeth: When blunt are called - obtuse; long and sharp called Acute - Fang like – caniniform

Conus Arteriosus: A structure in fish hearts that connects the ventricle to the ventral aorta and contains valves to prevent backflow.

Copulation: The act of sexual intercourse between individuals, typically leading to fertilization.

Corpuscles of Stannius: Structures associated with the kidneys of certain fish species that regulate calcium balance.

Cranial Nerves: Ten pairs of nerves originating from the brain, each serving specific functions related to sensory and motor activities.

Cross Section: A cut made perpendicular to the long axis of a structure, allowing for examination of internal features.

Ctenoid: With spiny (comb like) hind margin - refers to scales

Depth: Vertical height of body

Dermis: The inner layer of skin in fish, containing blood vessels, nerves, sense organs, and connective tissue, contributing to scale formation and structural support.

Dorsal: Pertaining to the back

Emarginate: With definitely the margin slightly hallowed; Notched forked. Shallow notch in the tail

Endocrine Gland: A gland that secretes hormones directly into the bloodstream or surrounding tissues to regulate physiological processes.

Endocrine System: A system of glands that produce and secrete hormones to regulate various bodily functions.

Epidermis: The outermost layer of skin in fish, consisting of flattened epithelial cells responsible for protection and pigment production.

Filamentous: Thread - like, filiform

Filiform: An extremely elongated and slender body shape resembling a thread

Fin Rays: Bony structures supporting the fins of fish, also used in age determination.

Fin Spines: Rigid, spiny structures found in the fins of some fish species, used in age determination similar to fin rays.

Finlet: Small fins in series behind the dorsal fin and ventral; Division of fin into series of smaller units

Forked: Connected basally but separated distally.

Frontal Plane: An anatomical plane that divides the body into dorsal and ventral halves, running perpendicular to the sagittal plane and parallel to the anterior-posterior axis.

Gill arch: The skeleton supporting gills

Gill Arch: A structural component of the gills, consisting of a cartilaginous or bony arch supporting the gill filaments and lamellae.

Gill Filaments (Primary Gill Lamellae): Blade-like structures supported by gill arches, increasing the surface area available for gas exchange in the gills.

Gill Lamellae (Secondary Gill Lamellae): Thin, semicircular structures lining the gill filaments, where the actual exchange of gases occurs between the blood and surrounding water.

Gill rakers: A series of bony projection along the anterior edge of the gill-arches

Gill Rakers: Comb-like structures found on the inner margin of each gill arch, involved in filtering water and preventing the entry of debris and food particles into the gill filaments.

Gill Slits: External openings in the body of fish, including sharks and rays, through which water exits after passing over the gills during respiration.

Gills: Respiratory organs in aquatic animals, including fish, that facilitate the exchange of oxygen and carbon dioxide between the bloodstream and the surrounding water.

Globiform: A body shape resembling a globe or sphere

Gonads: Reproductive organs responsible for producing gametes (sperm or eggs).

Growth Checks: Markers on hard parts of fish, such as scales or otoliths, used to determine age.

Heterocercal: Caudal fin with longer upper lobe

Incisors: Flattened front cutting teeth, which is chisel shaped

Interrenal Tissue (Adrenocortical Tissue): Tissue associated with the adrenal glands that secretes hormones such as cortisol.

Intestinal Mucosa: The lining of the intestines that produces hormones involved in regulating digestion and metabolism.

Iris: The round, pigmented membrane surrounding the pupil of the eye

Lunate: Shaped like a crescent

Mandible: The bone or bones forming the lower jaw

Maxilla: The principal side bone of the upper jaw

Maxillary: Pertaining to the upper jaw

Median fins: Combined terms of dorsal, caudal and ventral fins

Micropyle: A small opening in the chorion through which sperm enters the egg during fertilization.

Molar: A blunt and rounded grinding tooth

Mosaic or Determinate Development: Embryonic development where each blastomere formed during cleavage has limited potential to develop into a complete embryo.

Nerve Cells: Responsible for receiving and transmitting impulses.

Neuron: The fundamental and functional unit of the nervous system, responsible for transmitting electrical impulses and facilitating communication between various parts of the body.

Neurotransmitters: Chemical substances stored in synaptic vesicles within the presynaptic knob and include Acetylcholine (Ach), Norepinephrine (NE), Dopamine (DA), Gamma-Aminobutyric acid (GABA), Serotonin, Glutamic acid, and Glycine.

Notochord: A flexible rod-like structure present in the embryos of all chordates and in the adults of some primitive chordates. In fishes, it provides support and acts as a precursor to the vertebral column.

Osteichthyes: A class of bony fishes, characterized by a skeleton primarily composed of bone rather than cartilage. Fusiform: A torpedo-shaped body form observed in many fast-swimming fish species, characterized by a streamlined shape that reduces drag in the water.

Parental Care: Behaviours exhibited by parents to ensure the survival and well-being of their offspring.

Pericardial Membrane: A protective membrane surrounding the heart.

Plasma: The liquid component of blood, consisting of water, ions, proteins, and other solutes.

Primary Sexual Characters: Sex organs (testis and ovary) directly involved in the reproductive process.

Regulative or Indeterminate Development: Embryonic development where portions of the egg can give rise to separate organisms, allowing for compensation if some parts are removed.

Scales: Bony or horny plates covering the skin of fish

Secondary Sexual Characters: Morphological features not directly related to reproduction but assist in behaviours associated with mating, such as recognition of the opposite sex, copulation, and parental care.

Sexual Dimorphism: The morphological differences between male and female individuals of a species.

Single-cycle Circulation: Circulation pattern in which blood passes through the heart only once during a complete circuit through the body.

Sinus Venosus: A thin-walled sac that collects blood from veins before it enters the atrium.

Spinal Nerves: Paired nerves arising segmentally from the spinal cord and carrying sensory and motor fibres from and to various parts of the body.

Spiracle: A respiration opening behind the eye in sharks and rays

Spleen: An organ involved in the production and storage of blood cells, especially in sharks and rays.

Supporting Cells: Provide structural and functional support to nerve cells.

Ultimobranchial Tissue: Tissue associated with the thyroid gland that produces the hormone calcitonin.

Ventricle: The lower chamber of the heart that pumps blood out to the body.

Vertebrae: Bones forming the spinal column

Villiform teeth: Small s lender teeth forming velvety bands

Vomer: A bone forming the front part of the roof of the mouth

Zygote: The cell formed by the fusion of a sperm cell and an egg cell during fertilization.

References

B.C. Jhingran (1982) Fish and Fisheries

Biswas, S.P. (2002). Fundamentals of Ichthyology Narendra Publishing House, Delhi

De Vlaming, V. L. (1972). Environmental and endocrine control of teleost reproduction. Fish Physiology, 11, 13-83

Goodrich E. S. (1930). Studies on the structure and Development of vertebrates

H. R. Singh (1994). Advances in Fish Biology

Howard, J., 2019. *Fish biology and fisheries. Scientific e-Resources.* Jones F.R.H (1968). Fish Migration

K. C. Badapanda, (2012). Fishery Biology Narendra Publishing House, Delhi Kyle H. (1926). The Biology of fishes

Lowerre-Barbieri, S. K., Ganias, K., Saborido-Rey, F., Murua, H., & Hunter, J. R. (2011). Reproductive timing in marine fishes: variability, temporal scales, and methods. Marine and Coastal Fisheries, 3(1), 71-91.

Moyle P. B & Cech J. J (1998). Fishes – An Introduction to Ichthyology

N.S. Pillai & S. Paramasivam (2004) Introduction to Fishery Science.

Radha, v. and Altaff, k., 2023. Anatomy and histological evaluation of the reproductive system of marine calanoid copepod Centropages furcatus from a mass culture perspective. Aquatic Research, 6(4), pp.233-244.

S. Ayyappan, K.K. Vass, and G.V. Reddy (2007). Inland Fishery Resources of India: Their Potential and Management

S. S. Khanna & H. R. Singh (2002). A textbook of Fish Biology and Fisheries Narendra Publishing House, Delhi

Tony J. Pitcher (2012) The Behaviour of Teleost Fishes

U.K. Sarkar and A. K. Singh (2000). Inland Fisheries: Ecology and Management

Wootton, R. J. (1990). Ecology of teleost fishes.